Edge Computing

IEEE Press
445 Hoes Lane
Piscataway, NJ 08854

Edge Computing

Systems and Applications

Lanyu Xu
Department of Computer Science and Engineering
Oakland University, Rochester
Michigan, United States

Weisong Shi
Department of Computer and Information Sciences
University of Delaware, Newark
Delaware, United States

IEEE*Press*
WILEY

Published by John Wiley & Sons, Inc., Hoboken, New Jersey.
Published simultaneously in Canada.

Library of Congress Cataloging-in-Publication Data Applied for

Hardback ISBN: 9781394285839

Cover Design: Wiley
Cover Image: © VectorMine/stock.adobe.com

Set in 9.5/12.5pt STIXTwoText by Straive, Chennai, India

Contents

About the Authors *xiii*
Preface *xv*
About the Companion Website *xvii*

1 **Why Do We Need Edge Computing?** *1*
1.1 The Background of the Emergence *1*
1.2 The Evolutionary History *6*
1.2.1 Technology Preparation Period *7*
1.2.1.1 Distributed Database Models *7*
1.2.1.2 Peer-to-Peer (P2P) Computing Models *8*
1.2.1.3 Content Delivery Network (CDN) Models *9*
1.2.1.4 Function Cache and Cloudlet *10*
1.2.1.5 Mobile Edge Computing *10*
1.2.1.6 Fog Computing *11*
1.2.1.7 Cloud-Sea Computing *12*
1.2.2 Rapid Growth Period *12*
1.2.3 Intelligence Integration Period *14*
1.3 What Is Edge Computing? *15*
1.4 Summary and Practice *18*
1.4.1 Summary *18*
1.4.2 Practice Questions *18*
1.4.3 Course Projects *18*
Chapter 1 Suggested Papers *19*
References *19*

2 **Fundamentals of Edge Computing** *23*
2.1 Distributed Computing *23*
2.1.1 Distributed Computing Technologies *24*
2.1.2 Distributed System Platforms *25*

2.2 The Basic Concept and Key Characteristics of Edge Computing *26*
2.2.1 The Basic Concept *27*
2.2.2 The Key Characteristics *29*
2.2.2.1 Compute Migration *29*
2.2.2.2 5G and 6G Communication Technologies *30*
2.2.2.3 Advanced Storage Systems *31*
2.2.2.4 Lightweight Libraries and Kernels *32*
2.2.2.5 The Edge Computing Programming Model *32*
2.3 Edge Computing vs. Cloud Computing *33*
2.3.1 The Concept of Cloud Computing *34*
2.3.2 The Big Data Era *35*
2.3.3 Edge Computing vs. Cloud Computing *36*
2.3.4 Advantages and Challenges of Edge Computing *39*
2.4 Summary and Practice *42*
2.4.1 Summary *42*
2.4.2 Practice Questions *42*
2.4.3 Course Projects *42*
Chapter 2 Suggested Papers *43*
References *43*

3 Architecture and Components of Edge Computing *47*
3.1 Edge Infrastructure *47*
3.1.1 Introduction to Edge Computing Architecture *47*
3.1.1.1 Differentiation from Centralized Models *48*
3.1.2 Different Grades/Layers of Edge *49*
3.1.2.1 On-premises Edge *49*
3.1.2.2 Network Edge *50*
3.1.2.3 Data Center Edge *51*
3.1.3 Capabilities of Edge Infrastructure *51*
3.1.3.1 Data Process *51*
3.1.3.2 Cache and Storage *52*
3.1.3.3 Communication *52*
3.1.3.4 CDNs *52*
3.1.4 New Progress of Edge Computing Architecture *53*
3.1.4.1 Edge Collaborative Consortium Architecture *53*
3.1.4.2 Computing-Networking Integration Architecture *53*
3.1.4.3 Edge-Native Architecture *53*
3.1.5 Open Questions *54*
3.1.5.1 Computing Ability *54*
3.1.5.2 Programmability *54*
3.2 Edge Computing Models *55*

3.2.1 Overview and Definitions *55*
3.2.2 Collaborative Edge Computing Models *57*
3.2.2.1 Edge-to-Edge Collaboration *57*
3.2.2.2 Edge-to-Device Collaboration *58*
3.2.2.3 Edge-to-Cloud Collaboration *59*
3.2.2.4 Cloud-Edge-Device Collaboration *60*
3.2.3 Choosing the Right Model *61*
3.2.3.1 Factors to Consider *61*
3.2.3.2 Implementation Challenges and Solutions *62*
3.2.3.3 Typical Edge Computing Systems and Models *62*
3.2.3.4 New Progress of Edge Model *63*
3.2.4 Open Questions *64*
3.2.4.1 Continuity *64*
3.2.4.2 Unity *64*
3.3 Networking in Edge Computing *65*
3.3.1 Introduction and Development Process of Edge Computing-Network Integration *65*
3.3.1.1 The Development Process *67*
3.3.2 Edge Computing-Network Architectures *68*
3.3.2.1 Edge Network Design *68*
3.3.2.2 Distributed Networking *68*
3.3.2.3 Top-Down, Service-Enabled Convergence Architectures *69*
3.3.2.4 Computing Power-Aware Architectures *69*
3.3.3 Current Progress and Future Trend *69*
3.4 Summary and Practice *71*
3.4.1 Summary *71*
3.4.2 Practice Questions *71*
3.4.3 Course Projects *71*
Chapter 3 Suggested Papers *72*
References *72*

4 Toward Edge Intelligence *77*
4.1 What Is Edge Intelligence? *77*
4.1.1 Formal Definition *78*
4.2 Hardware and Software Support *80*
4.2.1 Hardware *81*
4.2.1.1 Application-Specific Integrated Circuit (ASIC) *81*
4.2.1.2 Field Programmable Gate Arrays (FPGAs) and Graphics Processing Units (GPUs) *84*
4.2.2 Software *86*
4.2.2.1 Packages *86*

4.2.2.2 Running Environment *88*
4.2.3 Container *90*
4.3 Technologies Enabling Edge Intelligence *91*
4.3.1 Compression Techniques *91*
4.3.1.1 Parameter Sharing and Pruning *91*
4.3.1.2 Quantization *94*
4.3.1.3 Low-Rank Approximation *97*
4.3.1.4 Knowledge Distillation *98*
4.3.1.5 Combined Compression Techniques *100*
4.3.2 Hardware-Software Codesign for Edge Optimization *101*
4.3.3 Applying Deep Learning Models on Resource-Constrained Edges *102*
4.4 Edge Intelligent System Design and Optimization *104*
4.4.1 Training on Edge *104*
4.4.1.1 Architecture for Model Training on Edge *105*
4.4.1.2 Optimization for Training *105*
4.4.1.3 Collaborative Training *106*
4.4.2 Model Inference on Edge *107*
4.4.2.1 Model Design *107*
4.4.2.2 Efficient AI *107*
4.4.2.3 Optimization Tool *108*
4.4.2.4 Collaborative Inference *108*
4.5 Summary and Practice *111*
4.5.1 Summary *111*
4.5.2 Practice Questions *112*
4.5.3 Course Projects *112*
Chapter 4 Suggested Papers *113*
References *113*

5 Challenges and Solutions in Edge Computing *123*
5.1 Programmability and Data Management *123*
5.1.1 Programmability *123*
5.1.2 Automatic Program Partitioning *125*
5.1.3 Naming Conventions *126*
5.1.4 Data Abstraction *128*
5.2 Resource Allocation and Optimization *130*
5.2.1 Scheduling Strategies *130*
5.2.2 Data Offloading and Load Balancing *131*
5.2.2.1 Data Offloading *132*
5.2.2.2 Load Balancing *133*
5.2.3 Optimization Metrics *133*
5.2.3.1 Latency *134*

5.2.3.2 Bandwidth *134*
5.2.3.3 Energy Consumption *135*
5.2.3.4 Cost *135*
5.3 Security, Privacy, and Service Management *136*
5.3.1 Privacy Protection and Security *136*
5.3.1.1 Sensor Security *137*
5.3.1.2 Securing Edge Networks and Platforms *137*
5.3.1.3 Data Sharing Security *138*
5.3.2 Edge Service Management *140*
5.3.2.1 Differentiation *140*
5.3.2.2 Extensibility *140*
5.3.2.3 Isolation *140*
5.3.2.4 Reliability *141*
5.4 Deployment Strategies and Integration *142*
5.4.1 Edge Nodes Deployment *142*
5.4.1.1 Selecting Edge Nodes *142*
5.4.1.2 Choosing Edge Data *143*
5.4.1.3 Ensuring Edge Node Reliability *143*
5.4.2 Deployment of AI Models on Resource-Constrained Edge Devices *143*
5.4.2.1 Edge Intelligence Devices *144*
5.4.2.2 Edge Intelligence Models *144*
5.4.3 Integration with Vertical Industries *145*
5.4.3.1 Bridging the Gap with Industry Standards *145*
5.4.3.2 Enhancing Data Protection and Access Mechanisms *145*
5.4.3.3 Improving Interoperability *145*
5.4.4 Hardware and Software Selection *146*
5.5 Foundations and Business Models *147*
5.5.1 Theoretical Foundations *147*
5.5.2 Business Models *148*
5.6 Summary and Practice *149*
5.6.1 Summary *149*
5.6.2 Practice Questions *151*
5.6.3 Course Projects *151*
Chapter 5 Suggested Papers *152*
References *152*

6 Future Trends and Emerging Technologies *157*
6.1 Edge Computing and New Paradigm *157*
6.1.1 Related New Paradigms *157*
6.1.1.1 Sky Computing *158*

6.1.1.2 Computing Power Network *160*
6.1.1.3 Meta Computing *161*
6.1.2 What Is New for Edge Computing *163*
6.1.3 Future *164*
6.2 Integration with Artificial Intelligence *164*
6.2.1 Basic Overview and Why Need Edge Computing *165*
6.2.2 Integrating LLM with Edge Computing *167*
6.2.2.1 Training with Edge Computing *167*
6.2.2.2 Inferring with Edge Computing *168*
6.2.2.3 Model Caching with Edge Computing *170*
6.2.3 Integration with Generative AI *171*
6.2.4 Applications and Future *172*
6.3 6G and Edge Computing *174*
6.3.1 Basic Understanding for 6G *174*
6.3.2 Mutual Influence: 6G and Edge Computing *175*
6.3.2.1 Edge Computing-Enabled 6G *175*
6.3.2.2 6G-Supported Edge Computing *177*
6.3.3 Potential Applications and Challenges *179*
6.4 Edge Computing in Space Exploration *180*
6.4.1 Basic Concepts *180*
6.4.2 Advanced Concepts and Architecture *182*
6.4.2.1 Advantages of Orbital Edge Computing *182*
6.4.2.2 Typical Architectures in OEC *183*
6.4.3 Advanced Scenarios and Challenges *184*
6.5 Summary and Practice *186*
6.5.1 Summary *186*
6.5.2 Practice Questions *186*
6.5.3 Course Projects *187*
Chapter 6 Suggested Papers *187*
References *188*

7 Case Studies and Practical Applications *193*
7.1 Manufacturing *195*
7.2 Telecommunications *198*
7.3 Healthcare *200*
7.4 Smart Cities *203*
7.5 Internet of Things *210*
7.6 Retail *211*
7.7 Autonomous Vehicles *213*

7.8 Summary and Practice *217*
7.8.1 Summary *217*
7.8.2 Practice Questions *217*
7.8.3 Course Projects *218*
Chapter 7 Suggested Papers *218*
References *218*

8 Privacy and Bias in Edge Computing *223*
8.1 Privacy in Edge Computing *224*
8.1.1 Privacy Concerns at Edge Computing *224*
8.1.2 Various Forms of Privacy *225*
8.1.2.1 Data Privacy *226*
8.1.2.2 Location Privacy *226*
8.1.2.3 Identity Privacy *226*
8.1.3 Introduction of Privacy-Preserving Techniques *227*
8.1.3.1 Data Anonymization *227*
8.1.3.2 *k*-Anonymity *228*
8.1.3.3 Differential Privacy *230*
8.1.3.4 Privacy Budget *231*
8.1.3.5 Global Sensitivity *231*
8.1.3.6 Noise Addition *231*
8.1.3.7 The Laplacian Mechanism *231*
8.1.3.8 Homomorphic Encryption *232*
8.1.3.9 Federated Learning *234*
8.1.4 Open Research Problems *235*
8.2 Accessibility and Digital Divide *236*
8.2.1 What Is Bias? *236*
8.2.2 Types of Biases *237*
8.2.3 Causes of Biases? *241*
8.2.4 Bias Impact on Edge Computing Algorithms *242*
8.2.5 Bias Mitigation Techniques *243*
8.2.5.1 Bias Mitigation in Algorithms *243*
8.2.5.2 Preprocessing and In-processing *243*
8.2.5.3 In-processing *244*
8.2.5.4 Postprocessing *245*
8.2.6 Open Research Problems *245*
8.3 Summary and Practice *245*
8.3.1 Summary *245*
8.3.2 Practice Questions *246*

8.3.3 Course Projects *246*
Chapter 8 Suggested Papers *247*
References *247*

9 Conclusion and Future Directions *253*
9.1 Key Insights and Conclusions *253*
9.2 So, What Is Next? *254*

Index *257*

About the Authors

Lanyu Xu is currently an Assistant Professor of Computer Science and Engineering at Oakland University. Dr. Xu leads the Edge Intelligence Systems Lab. Her research intersects edge computing and deep learning, emphasizing the development of efficient edge intelligence systems. Her work explores optimization frameworks, intelligent systems, and AI applications to address challenges in efficiency and real-world applicability of edge systems across various domains.

Weisong Shi is an Alumni Distinguished Professor and Department Chair of Computer and Information Sciences at the University of Delaware (UD). Dr. Shi leads the Connected and Autonomous Research (CAR) Laboratory. He is an internationally renowned expert in edge computing, autonomous driving, and connected health. His pioneer paper, "Edge Computing: Vision and Challenges," has been cited over 8000 times. He is the Editor-in-Chief of *IEEE Internet Computing Magazine* and the founding steering committee chair of several conferences, including the ACM/IEEE Symposium on Edge Computing (SEC), IEEE/ACM International Conference on Connected Health (CHASE), and the IEEE International Conference on Mobility (MOST). He is a fellow of IEEE and a distinguished member of Association for Computing Machinery (ACM).

About the Authors

Preface

Over the past decade, edge computing's rapid evolution has fundamentally transformed how data is processed, stored, and utilized across multiple industry sectors, such as smart manufacturing, healthcare, smart cities, and transportation. As a critical enabler of technologies such as the Internet of Things (IoT), autonomous systems, and real-time analytics, edge computing has progressed from a nascent concept to widespread adoption. Despite this remarkable growth, there remains a lack of educational resources dedicated to equipping the next generation of the workforce with the knowledge and skills needed to advance edge computing further.

This book was motivated to address that gap, offering a comprehensive introduction to edge computing's principles, architectures, applications, and challenges. It aims to provide readers, ranging from students to professionals, with a solid foundation in edge computing, enabling them to understand its current state, tackle its challenges, and drive its development. By bridging theory and practice, this book aspires to inspire innovation, foster collaboration, and promote growth in this rapidly evolving field.

Designed as both a textbook and a reference guide, this book includes practice questions, course projects, and curated reading materials for each chapter to enhance learning. Readers with diverse interests and goals can navigate directly to the chapters most relevant to them, making the book a flexible resource for students, educators, researchers, and professionals alike.

The book is structured into nine chapters. Chapter 1 introduces the importance of edge computing, providing its background and evolutionary history. Chapter 2 lays the groundwork, covering fundamental principles, models, and technologies that underpin edge computing. Chapter 3 delves into the architecture and components of edge computing, including infrastructure and collaborative models. Chapter 4 transitions into edge intelligence by highlighting the integration of artificial intelligence with edge computing. Chapter 5 addresses key challenges such as programmability, resource optimization, and security, while proposing potential solutions. Chapter 6 looks to the future, discussing emerging paradigms like sky computing, 6G, and edge computing in space exploration. Chapter 7 provides practical insights through real-world case studies, illustrating edge computing's impact on industries such as manufacturing, healthcare, smart cities, and more. Chapter 8 examines privacy concerns and the digital divide, exploring biases, their impacts, and mitigation strategies in edge computing. Chapter 9 concludes the book.

This endeavor would not have been possible without the unwavering dedication and expertise of the team behind this book. We are deeply grateful to the contributors, editors, and reviewers whose insights and hard work shaped this book. In particular, we extend our heartfelt thanks to Dr. Haihua Chen (University of North Texas), Dr. Shihong Hu (Hohai University), Dr. Sidi Lu (William & Marry), Dr. Kewei Sha (University of North Texas), Dr. Qingyang Zhang (Anhui University), Dr. Xingzhou Zhang (Chinese Academy of Science), and PhD students Komala Subramanyam Cherukuri (University of North Texas), Yuankai He (University of Delaware), Shaibal Saha (Oakland University), Qiren Wang (University of Delaware), Yichen Xia (University of Delaware), and Yongtao Yao (University of Delaware).

To all the readers, we hope you enjoy reading the book and find the book serves as both a resource and an inspiration as you explore the exciting world of edge computing.

January, 2025

Lanyu Xu
Rochester, United States

Weisong Shi
Newark, United States

About the Companion Website

This book is accompanied by a companion website:

www.wiley.com/go/xu/computing

This website contains the PowerPoint slides for each chapter.

1

Why Do We Need Edge Computing?

What is edge computing? Why did it become popular after being proposed? What are the relationships between edge computing and IoT/Cloud Computing? In this chapter, we will answer these three questions by introducing the background, the evolutionary history, and the concept of edge computing.

1.1 The Background of the Emergence

To answer the question of this chapter, let us trace back to when edge computing was proposed, back to the big data era when the Internet of Things (IoT) and cloud computing were blooming.

The IoT technology [3] aims to connect physical objects to the Internet according to the communication protocols of IoT, utilizing technologies such as RFID (radio frequency identification), wireless data communication, and GPS (global positioning system). This enables information exchange for intelligent identification, positioning, tracking, monitoring, and management of Internet resources. IoT has significantly expanded with the advancement of computer and network communication technologies. It now encompasses the integration of almost all information technologies with computer and network technologies, facilitating real-time data sharing between objects and achieving intelligent real-time data collection, transmission, processing, and execution. The concept of "computer information perception without human intervention" has gradually been applied to fields such as wearable devices, smart homes, environmental sensing, intelligent transportation systems, and smart manufacturing [18, 36]. Key technologies involved in IoT include:

- **Sensor Technology**: This involves acquiring information from natural sources, processing (transforming), and identifying it. Sensor technology is a critical

Edge Computing: Systems and Applications, First Edition. Lanyu Xu and Weisong Shi.

Companion Website: www.wiley.com/go/xu/computing

aspect of computer applications, as it senses (or responds to) and detects specific information from the measured object, converting it into output signals according to certain rules.

- **RFID Technology**: This comprehensive technology integrates radio frequency and embedded technologies to automatically identify target objects and obtain related data through radio frequency signals. The identification process does not require human intervention and can operate in various harsh environments, with promising and broad applications in automatic identification, logistics management, and more.
- **Embedded System Technology**: This is a complex technology that integrates computer hardware and software, sensor technology, integrated circuit technology, and electronic application technology. Over the decades, intelligent terminal products characterized by embedded systems have become ubiquitous, ranging from smartwatches to aerospace satellite systems. Embedded systems are transforming people's lives, driving industrial production, and advancing the defense industry. If we make a simple analogy of the IoT to the human body, sensors are akin to human senses like eyes, nose, and skin; the network is the nervous system transmitting information, and the embedded system is the brain that classifies and processes the received information.

Later on, with the rapid development of IoT and the widespread adoption of 4G/5G wireless networks, the era of the Internet of Everything (IoE) [11] has arrived. Cisco introduced the concept of IoE in December 2012. It represents a new network architecture for future Internet connectivity and the evolution of IoT, enhancing the network's intelligent processing and security features. IoE employs a distributed structure, integrating application-centric networks, computing, and storage on a new platform. It is driven by IP settings, global higher bandwidth access, and IPv6, supporting hundreds of millions of edge terminals and devices connected to the Internet. Compared to IoT, IoE not only involves "thing-to-thing" connections, but also introduces a higher level of "human-to-thing" connectivity. Its distinguishing feature is that any "thing" will possess contextual awareness, enhanced computing capabilities, and sensing abilities.

Integrating humans and information into the Internet, the network will have billions or even trillions of connected nodes. The IoE is built on the physical network, enhancing network intelligence to achieve integration, coordination, and personalization among the "things" on the internet.

Application services based on the IoE platform require shorter response times and will generate a large amount of data involving personal privacy. For example, sensors and cameras installed on autonomous vehicles capture road condition information in real time; one car with five cameras can generate more than 24 terabytes (TB) data per day [17]. According to the Insurance Institute for Highway Safety, there will be 3.5 million self-driving vehicles on U.S. roads by 2025 and 4.5 million by 2030 [21]. The Boeing-787 generates about 5 gigabytes (GB) of data per second and requires real-time processing of the data. In Beijing, China, the electric vehicle monitoring platform can provide continuous 7 × 24-hour real-time monitoring for 10,000 electric vehicles and forward data to various enterprise platforms at a rate of one data point every 10 seconds per vehicle. In terms of social security, the United States has deployed over 30 million surveillance cameras, generating more than 4 billion hours of video data each week. China's "Skynet" surveillance network, used for crime prevention, has installed over 20 million high-definition surveillance cameras nationwide, monitoring and recording pedestrians and vehicles in real time.

Since the concept was proposed in 2005, cloud computing has been widely applied, changing how people work and live. SaaS (Software as a Service) is commonly used in data centers of major IT companies like Google, Twitter, Facebook, and Baidu. Scalable infrastructure and processing engines supporting cloud services have significantly impacted application services such as Google File System (GFS), MapReduce programming model, Hadoop (a distributed system developed by Apache Foundation), and Spark (the in-memory computing framework designed by the AMP Lab at the University of California Berkeley). However, in the context of IoT and similar applications, data is geographically dispersed and demands higher response times and security. Although cloud computing provides an efficient platform for big data processing, the network bandwidth growth rate cannot keep up with the data growth rate. The cost reduction rate of network bandwidth is much slower than that of hardware resources like CPU and memory, and the complex network environment makes it challenging to significantly improve network latency. Therefore, the traditional cloud computing model will struggle to support application services based on IoE efficiently and in real time, requiring solutions to address the bandwidth and latency bottlenecks.

With the rapid development and widespread application of the IoE, edge devices are transitioning from primarily serving as data consumers to serving as both data producers and consumers. Simultaneously, network edge devices are gradually capable of utilizing the collected real-time data for pattern recognition, predictive analysis or optimization, and intelligent processing. In the edge computing model, computing resources are closer to the data source, and network edge devices now have sufficient computational power to process the raw data locally and send the results to the cloud computing center locally. The edge computing model not only reduces the bandwidth pressure in network transmission, speeding up data analysis and processing, but also lowers the risk of privacy leaks for sensitive terminal data.

Currently, big data processing is shifting from the centralized processing era centered on cloud computing (we refer to the years from 2005 to 2015 as the centralized big data processing era) to the edge computing era centered on the IoE (we refer to it as the edge-based big data processing era). During the centralized big data processing era, the focus was more on centralized storage and processing of big data, achieved by building cloud computing centers and leveraging their powerful computing capabilities to solve computational and storage issues centrally. In contrast, in the edge-based big data processing era, network edge devices generate massive real-time data. In 2018, Cisco's Global Cloud Index estimated that nearly 850 zettabytes (ZB) will be generated by all people, machines, and things by 2021. Yet only around 10% is classed as useful data; useful data is predicted to four times exceed data center traffic (21 ZB per year) [10]. From 2018 to 2023, the average number of devices owned per person worldwide increased from 2.4 to 3.6. Specifically, in North America, on average, one person owned eight devices in 2018 and 13 devices in 2023 [38]. According to Statista, the number of IoT devices connected to the network was 15.14 billion in 2023 and will reach 29.42 billion in 2030 [35]. This mismatch between data producing and data consuming requires the emergence of an alternation for cloud-based data centers. Instead of purely relying on cloud computing, data can be stored, processed, and analyzed at the network edge. These edge devices will be deployed on edge computing platforms supporting real-time data processing, providing users with numerous service or function interfaces, which users can invoke to obtain the necessary edge computing services.

Therefore, the linearly growing centralized cloud computing capacity can no longer match the exponential growth of massive edge data. Single computing resources based on the cloud computing model can no longer meet the demands for real-time processing, security, and low energy consumption in big data processing. Based on the existing centralized big data processing centered on the cloud computing model, there is an urgent need for edge big data processing technology centered on the edge computing model to handle the vast edge data. The two complement each other, applied to big data processing at both the cloud center and the edge end, addressing the inadequacies of cloud computing services in the IoE era.

When observing the data explosion in three dimensions: velocity, variety, and volume, we will find that the emergence and rapid development of edge computing is inevitable (Figure 1.1). The cloud-centralized computing paradigm performs well when the data is generated with a confined speed, size, and format. While in the IoE era, data is increasingly produced at the network's edge regarding velocity, variety, and volume. In terms of variety, different types of data (e.g., text, audio, and photo) are generated every day and every second from numerous devices (e.g., IoT, web browser, camera, and social media). These data are generated with different

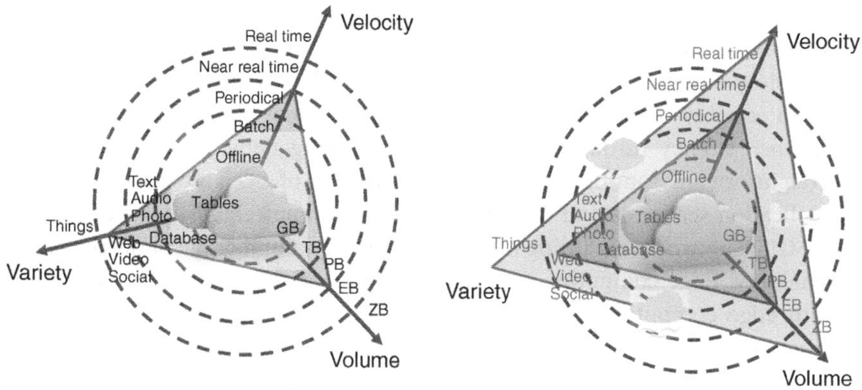

Figure 1.1 The increase of data pushes the evolution of edge computing.

velocities (e.g., real time, near real time, periodic, and batch generated). Therefore, storage and process requirements for these data will be different. With the tremendous number of devices and the frequent speed of data generation, there is no surprise that the volume of data generated is increasing dramatically. Megabytes (MB) and TB have become the typical units. In fact, as of 2024, the amount of data generated per day is around 328.77 million TB, which equals 0.33 ZB [13]. Given these factors, relying purely on the cloud for all data processing is impossible. This is not just because of the computing pressure brought to the cloud computation center but also because the bandwidth capability required for transmitting this amount of data is challenging. The only solution to process the data reliably and in a timely manner is edge computing, which ensures a shorter response time, more efficient processing, and smaller network pressure.

Compared to cloud computing, edge computing can better support mobile computing and IoT applications, offering the following distinct advantages:

- **Greatly Alleviates Network Bandwidth and Data Center Pressure**: With the development of IoT, global devices will generate massive amounts of data. However, only a small portion of this data is critical, while most of it is temporary and does not need long-term storage (the amount of data generated by devices is two orders of magnitude higher than the amount of data that needs to be stored). Edge computing can fully utilize geographically distributed network edges to process a large amount of temporary data, thereby reducing the pressure on network bandwidth and data centers.
- **Enhances Service Responsiveness**: The inherent limitations of mobile devices in computing, storage, and power resources are evident. Cloud computing can provide services to mobile devices to address these deficiencies. However, network transmission speeds are constrained by the development of

communication technologies, and in complex network environments, issues such as unstable connections and routing further exacerbate latency, jitter, and slow data transmission speeds, severely affecting the responsiveness of cloud services. Edge computing offers services near the user, ensuring low network latency through proximity and reducing network jitter with more straightforward routing. With the development of 5G and 6G, the diverse application scenarios and differentiated service requirements pose challenges to 5G/6G networks regarding throughput, latency, number of connections, and reliability. Edge computing and 5G/6G technologies complement each other, with edge computing leveraging localization, proximity, and low latency to drive 5G/6G architectural changes, while 5G/6G technology is essential for reducing data transmission latency and enhancing service responsiveness in edge computing systems.

- **Protects Privacy Data and Enhances Data Security**: Data security has always been a critical issue in IoT applications. Surveys show that approximately 86% of the U.S. general population is concerned about data privacy [23]. In the cloud computing model, all data and applications are stored in data centers, making it difficult for users to have fine-grained control over data access and usage. Edge computing provides the infrastructure for storing and using critical privacy data, restricting the operation of privacy data within firewalls and thereby enhancing data security (for more detailed information, see Chapter 8).

1.2 The Evolutionary History

The field of edge computing has developed rapidly since 2014. We categorize the development process into three stages: technology preparation period, rapid growth period, and steady development period. We use "edge computing" as the keyword to search the number of articles published per year in Google Scholar. As shown in Figure 1.2, before 2015, edge computing was in the technology preparation period. Since 2015, the number of papers related to "edge computing" has grown tenfold. Edge computing has entered a rapid growth period. The number of papers has been increasing and reaching a steady development period since 2020. In this period, the development is focused on integrating academia and industry, bringing the product into the business, and finally facilitating peoples' daily lives. Figure 1.3 illustrates typical events in the development process of edge computing.

The development of edge computing is closely linked to the evolution of data-oriented computing models. As the scale of data increases, the demand for performance and energy efficiency in data processing continues to grow.

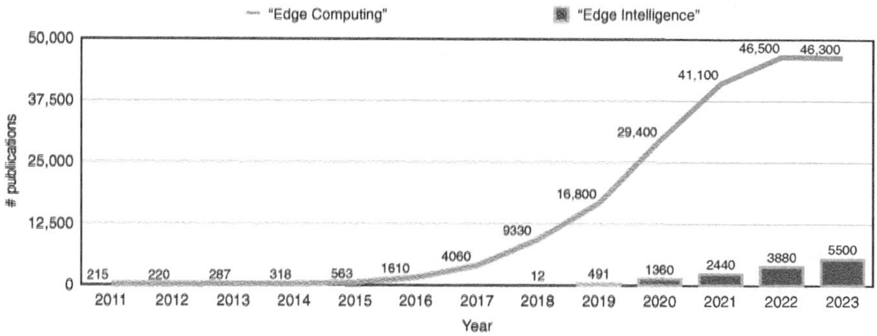

Figure 1.2 Number of publications searched by keyword "edge computing" and "edge intelligence."

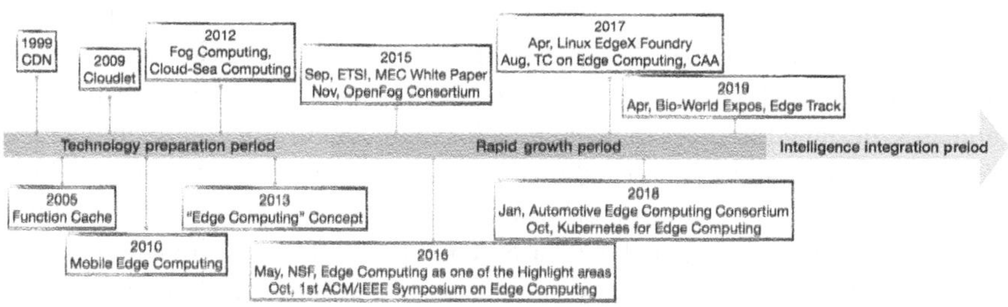

Figure 1.3 The evolution of edge computing and key milestones.

To address the issues of computational load and data transmission bandwidth in data transfer, computation, and storage processes, researchers explored ways to enhance data-processing capabilities near data sources even before the advent of edge computing. This involves shifting computational tasks from centralized computing centers to the network edge. The main typical models include distributed database models, peer-to-peer (P2P) computing models, content delivery network (CDN) models, mobile edge computing models, fog computing models, and cloud-sea computing models. We will explain these different models in the order of their emergence and also introduce the history of edge computing.

1.2.1 Technology Preparation Period

During the technology preparation period, edge computing went through the development process of dormancy, presentation, definition, and generalization.

1.2.1.1 Distributed Database Models

The distributed database model results from combining database technology and network technology. In the era of big data, the growth in the variety and quantity of data has made distributed databases a core technology for data storage and processing. Distributed databases are deployed on self-organizing network servers or dispersed across the Internet, enterprise networks, Internet, and other independent computers in self-organizing networks. Data is stored on multiple machines, and operations are not limited to a single machine but allow transactions to be executed across multiple machines to improve database access performance.

Distributed databases have become a core technology for big data processing. Based on their structure, distributed databases include homogeneous and heterogeneous systems. The former has database instances running in environments with the same software and hardware, featuring a single-access interface. The latter operates in environments where hardware, operating systems, database management systems, and data models vary. Based on the types of data processed, distributed databases mainly include relational (such as Structured Query Language, SQL), nonrelational (such as NoSQL), extensible markup language (XML)-based, and NewSQL distributed databases. Among these, NoSQL and NewSQL distributed databases are the most widely used [16]. NoSQL distributed databases, designed to meet the demands for high concurrency, efficient storage access, high reliability, and scalability in big data environments, are divided into key-value stores, column stores, document-oriented databases, and graph databases. NewSQL distributed databases, characterized by real-time processing, complex analysis, and fast querying, are relational distributed databases designed for massive data storage in big data environments, including Google Spanner, Clustrix, and VoltDB. SQL-distributed databases are relational distributed databases, with typical examples including Microsoft's and Oracle's distributed databases. XML-based distributed databases mainly store data in XML format and are essentially document-oriented, similar to NoSQL distributed databases [22].

Compared to edge computing models, distributed databases provide data storage in big data environments but pay less attention to the heterogeneous computing and storage capabilities of the devices they reside on, focusing mainly on achieving distributed data storage and sharing. Distributed database technology requires significant space and offers lower data privacy. For distributed transaction processing across multiple databases, data consistency technology is a major challenge for distributed databases [14]. In edge computing models, data resides on edge devices, offering higher privacy, reliability, and availability. In the era of the IoE, "heterogeneous edge architectures and the need to support multiple application services" will become the fundamental approach for edge computing models to handle big data processing.

1.2.1.2 Peer-to-Peer (P2P) Computing Models

P2P computing [27] is one of the early file transfer technologies that pushed computing to the edge of the network. The term *P2P* was first introduced in

2000 to implement file-sharing systems. Since then, it has gradually developed into an important subfield of distributed systems. The key research topics in P2P models include decentralization, maximizing scalability, tolerance of high-level node churn, and preventing malicious behavior. Major achievements in this field include:

- Distributed Hash Table (DHT), which later evolved into the general paradigm for key-value distributed storage in cloud computing models.
- Generalized gossip protocols, which have been widely used for complex task processing applications beyond simple information dissemination, such as data fusion and topology management.
- Multimedia streaming technology, in forms such as video on demand, real-time video, and personal communication.

However, widespread media coverage of P2P being used for illegal file sharing and related lawsuits has hindered the practical recognition of some commercial technologies based on the P2P model.

The edge computing model bears significant similarities to P2P technology, while it expands on the latter with new technologies and methods, extending the concept of P2P to network edge devices. This represents a fusion of P2P computing and cloud computing.

1.2.1.3 Content Delivery Network (CDN) Models

CDN [29] was proposed by Akamai in 1998. CDN is an Internet-based caching network, which relies on caching servers deployed in different places and points users' access to the nearest caching server through load balancing, content distribution, scheduling, and other functional modules of the central platform. Therefore, CDN can reduce network congestion and improve user access response speed and hit rate. It has gained significant attention from both academia and industry since it was proposed. Companies like Amazon [2] and Akamai [1] possess mature CDN technologies that provide users with the expected performance and experience while reducing the operational pressures on service providers.

Active content distribution networks (ACDNs), an improvement over traditional CDNs, help content providers avoid the hassle of predicting the preconfiguring resources and determining their locations [30]. ACDN allows applications to be deployed on any server and uses newly designed algorithms to replicate and migrate applications between servers as needed.

The concept of edge computing can be traced back to around the year 2000, when CDNs were deployed large scale. At that time, major companies like Akamai announced the distribution of web-based content through CDN edge servers. The primary goal of this method was to benefit from the short distances and available resources of CDNs to achieve large-scale scalability. In the early days of edge computing, the "edge" was limited to CDN cache servers distributed around the world. However, today's development of edge computing has far exceeded the scope of CDNs. The "edge" in the edge computing model is not confined to edge nodes; it includes any computational, storage, and networking resources along the path from data sources to cloud computing centers.

1.2.1.4 Function Cache and Cloudlet

To enable static content distribution, CDN emphasizes the backup and caching of data, while edge computing focuses more on function caching to improve computational capabilities. Function cache was proposed by Ravi et al. [31], where it is applied to personalized mailbox management services to save latency and bandwidth. Satyanarayanan et al. [33] introduced the concept of Cloudlet, which is a trusted, small-scale, and resource-rich host, located at the edge of the network, connected to the Internet, and can be accessed by mobile devices to provide services. Cloudlet is also known as "small cloud" as it can provide services for users, similar to the cloud server. At this point, edge computing focused on the downstream transfer of functions from cloud servers to edge servers, aiming to reduce bandwidth usage and minimize delays.

1.2.1.5 Mobile Edge Computing

The development of the IoE has enabled the interconnection of numerous types of devices, such as smartphones, tablets, wireless sensors, and wearable devices. However, the limited energy and computing resources of most network edge devices make the design of IoE particularly challenging. Mobile edge computing (MEC) [19] is a new network architecture that provides information technology services and cloud computing capabilities within the proximity of the mobile user's wireless access network. It has become a standardized and regulated technology. In 2014, the European Telecommunications Standards Institute (ETSI) introduced the standardization of the term MEC, highlighting that MEC provides a new ecosystem and value chain. Utilizing MEC, intensive mobile computing tasks can be offloaded to nearby network edge servers. Because MEC is located within the wireless access network and close to mobile users, it can achieve lower latency and higher bandwidth, thereby improving service quality and user experience. MEC is also a key technology in the development of 5G, helping to meet the high standards of 5G in terms of latency, programmability, and scalability. By deploying services and caches at the network edge, MEC reduces congestion in the core network and efficiently responds to use requests.

Task migration is one of the challenges in mobile computing technology, particularly in environments where continuous service availability and seamless user experience are crucial. The process involves transferring ongoing tasks from one computational node to another, which can be triggered by various factors such as device mobility, energy conservation needs, or load balancing requirements. Effective task migration must minimize latency, avoid data loss, and maintain application state continuity, which is challenging due to the heterogeneous and dynamic nature of mobile environments. Furthermore, ensuring security during data transfer, managing the energy consumption of mobile devices, and dealing with fluctuating network conditions are additional hurdles that need optimization solutions. As mobile computing continues to evolve, developing robust, efficient, and secure task migration mechanisms will be critical to fully leveraging the potential of mobile platforms. MEC has been applied in various scenarios, such as vehicular networks, IoT gateways, auxiliary computing, intelligent video acceleration, and mobile big data analysis.

MEC emphasized the establishment of edge servers between the cloud server and edge devices to process computing. However, mobile edge nodes are generally considered to lack computing capabilities. In contrast, the nodes in the edge computing model possess strong computing capabilities. Therefore, MEC resembles the architecture and hierarchy of an edge computing server, functioning as an important part of edge computing.

1.2.1.6 Fog Computing

Cisco introduced fog computing in 2012 and defined fog computing as a highly virtualized computing platform for migrating cloud computing center tasks to network edge devices [7]. Fog computing provides computing, storage, and network services between end devices and traditional cloud computing centers, complementing cloud computing. Vaquero and Rodero-Merino [39] have provided a comprehensive definition of fog computing, which extends cloud-based network architecture by introducing an intermediate layer between the cloud and mobile devices. This intermediate layer, known as the fog layer, consists of fog servers deployed at the network edge. Fog computing reduces the need for multiple communications between the cloud computing center and mobile users. It relieves the bandwidth load and energy consumption pressure of main links by reducing the number of communications between cloud computing centers and mobile users. When there is a large volume of mobile users, they can access cached content and request specific services from the fog computing servers. Additionally, fog computing servers can interconnect with cloud computing centers, leveraging their powerful computational capabilities and extensive applications and services.

The concepts of edge computing and fog computing have great similarities and often represent the same idea. If we are to distinguish between the two, this book posits that edge computing, in addition to focusing on infrastructure, also pays attention to edge devices and places more emphasis on the design and implementation of edge intelligence. In contrast, fog computing focuses more on the

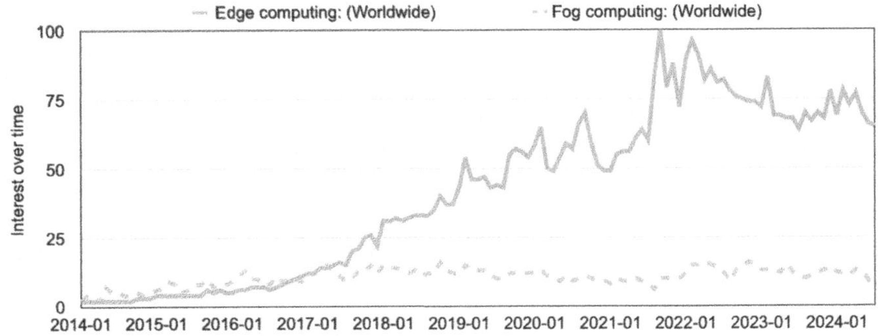

Figure 1.4 "Edge computing" and "fog computing" trends.

management of back-end distributed shared resources. As shown in Figure 1.4, since 2017, the level of attention to edge computing has gradually surpassed that of fog computing, and its attention continues to rise.

1.2.1.7 Cloud-Sea Computing

In the context of IoE, the amount of data to be processed will reach ZB levels. The sensing, transmission, storage, and processing capabilities of information systems need to be correspondingly enhanced. To address this challenge, in 2012, the Chinese Academy of Sciences launched a ten-year strategic priority research initiative called the Next Generation Information and Communication Technology (NICT) initiative. Its main purpose is to carry out research on the "Cloud-Sea Computing System Project" [40]. It aims to augment cloud computing by cooperation and integration of the "cloud computing" system and the "sea computing" system. "Sea" refers to an augmented client-side consisting of human-facing and physical world-facing devices and subsystems. The research focuses on proposing system-level solutions from perspectives such as overall system architecture, data center and server and storage system layers, and processor chip level.

Cloud-sea computing focuses on the two ends "sea" and "cloud" while edge computing focuses on the data path between "sea" and "cloud." Cloud-sea computing is a great subset example of edge computing.

1.2.2 Rapid Growth Period

Since 2015, edge commuting has been in a rapid growth period, attracting intensive close attention from academia and industry.

At the government level, in May 2016, the National Science Foundation (NSF) listed edge computing as one of the highlighted areas in the research of computer systems. In August 2016, NSF and Intel formed a partnership in information center networks in wireless edge networks (ICN-WEN) [37]. In October 2016, the NSF held the NSF Workshop on Grand Challenges in edge computing [8]. The workshop focused on three topics: the vision of edge computing in the next five to ten years, the grand challenges to achieving the vision, and the best mechanisms for academia, industry, and the government to attack these challenges in a cooperative way. This indicates that the development of edge computing has attracted great attention at the government level.

In academia, a formal definition of edge computing is given in the paper *Edge computing: vision and challenges* [34]. Edge computing is defined as enabling technologies that allow computation to be performed at the edge of the network, processing downstream data on behalf of cloud services, and upstream data on behalf of IoT services. This paper pointed out the challenges of edge computing and is one of the most cited papers in the edge computing field. In October 2016, ACM and IEEE jointly organized the first ACM/IEEE Symposium on Edge Computing (SEC). Since then, the International Conference on Distributed Computing Systems (ICDCS), the International Conference on Computer Communications (INFOCOM), the International Middleware Conference, and other important international conferences have added an edge computing track and/or workshops to their main conferences.

At the same time, multiple industry sectors have actively promoted the development of edge computing. In September 2015, ETSI published a white paper on MEC. In November 2015, Cisco, ARM, Dell, Intel, Microsoft, and Princeton University jointly established the OpenFog Consortium, which is dedicated to the development of Fog Reference Architecture [28]. The OpenFog Consortium merged into the Industrial Internet-of-Things (IIoT) in January 2019. In November 2016, Huawei, Shenyang Institute of Automation of Chinese Academy of Sciences, China Academy of Information and Communications Technology (CAICT), Intel, ARM, and iSoftStone established the Edge Computing Consortium (ECC) in Beijing, China, which is dedicated to advancing cooperation among industry resources from government, vendor, academic, research, and customer sectors, and pushing forward the sustainable development of the edge computing industry [12]. In March 2017, the ETSI MEC Industry Specification Working Group was formally renamed to multiaccess edge computing, aiming to better meet the requirements of edge computing and related standards. Linux EdgeX Foundry was also built in 2017; it is a vendor-neutral open-source project hosted by The Linux Foundation. It aims to build a common open framework for IoT edge computing. In January 2018, Automotive ECC (AECC) was established to drive the network and computing infrastructure needs of automotive big data [4], which indicates that edge computing is valued in the vehicle domain. In the same year, the Cloud Native Computing Foundation (CNCF) Foundation and Eclipse Foundation cooperated to bring Kubernetes, which has been widely used in the ultra large-scale cloud computing environment, into the edge computing scene of the IoT. Subsequently, KubeEdge, a Kubernetes native edge computing framework, was accepted into the CNCF sandbox in March 2019 [24]. In April 2019, the Bio-IT World Conference and Expos added the edge track [6], which means that edge computing is important to the health domain as well.

In the rapid growth period, the industry has seen significant advancements, evidenced by the availability of multiple edge environment solutions from major service providers. Today, options like AWS Greengrass [5], Microsoft Azure [25, 26], Google Cloud Platform Edge Zones [15] have made it easier for businesses and developers to deploy and manage edge computing infrastructures effectively. These developments underscore the maturity and widespread adoption of edge computing across various sectors.

1.2.3 Intelligence Integration Period

Edge computing has seen substantial growth and transformation in recent years, driven by the increasing demand for low-latency data processing and efficient resource utilization. The development of edge computing is characterized by bringing together IoT, big data, and mobile computing into an integrated and ubiquitous computing platform. The capability of delivering on-demand computing power at the edge and processing a vast amount of data from various devices/sensors enables real-time analytics and decision-making. A significant advancement within this domain is the integration of edge intelligence, where artificial intelligence (AI) and machine learning (ML) algorithms are deployed at the edge. This symbiotic relationship enhances the capability of edge computing, allowing for sophisticated data analysis and autonomous decision-making directly at the data source. Edge intelligence empowers devices to process and act on data locally, leading to smarter, faster, and more efficient systems across various industries, from autonomous vehicles to smart cities and beyond.

Building on this foundation, the intelligence integration period marks a crucial phase in the evolution of edge computing. Technological strategies such as pruning, quantization, and knowledge distillation are employed to optimize AI models for efficient operation on edge devices. Simultaneously, AI algorithms find wide application across systems such as smart surveillance, autonomous vehicles, health monitoring systems, industrial IoT, smart agriculture, and retail enhancements, further demonstrating the pervasive impact of this integration. These advancements not only improve responsiveness but also deliver substantial societal benefits. While the potential of these integrations is immense, the associated privacy and security concerns are non-negligible and will be discussed in Chapter 8, with deeper technological and application-based discussions slated for Chapters 3, 4, and 7.

1.3 What Is Edge Computing?

After exploring the background of edge computing's emergence and examining its three distinct phases of development, it's time to address a fundamental question: what exactly is edge computing?

There is no standard definition for edge computing yet. In the field of edge computing, industry experts and other researchers have provided their own definitions. For example, IBM views edge computing as a distributed computing framework that brings enterprise applications closer to data sources such as IoT devices or local edge servers [20]. CISCO interprets edge computing as a model that shifts computing resources from central data centers or public clouds closer to devices, that is, embedded at the edge of service provider networks [9]. Satyanarayanan defines edge computing as a computing paradigm in which substantial computing and storage resources—variously referred to as cloudlets, micro data centers, or fog nodes—are placed at the Internet's edge in close proximity to mobile devices or sensors [32]. Yousefpour et al. believe edge computing is located at the edge of the network close to IoT devices, and edge can be more than one hop away from IoT devices in the local IoT network [41].

Here, we give our own definition of edge computing. In the vision paper published in 2016, we highlighted that **edge computing refers to the enabling technologies allowing computation to be performed at the edge of the network, on downstream data on behalf of cloud services and upstream data on behalf of IoT services.** [34]. As shown in Figure 1.5, in our definition,

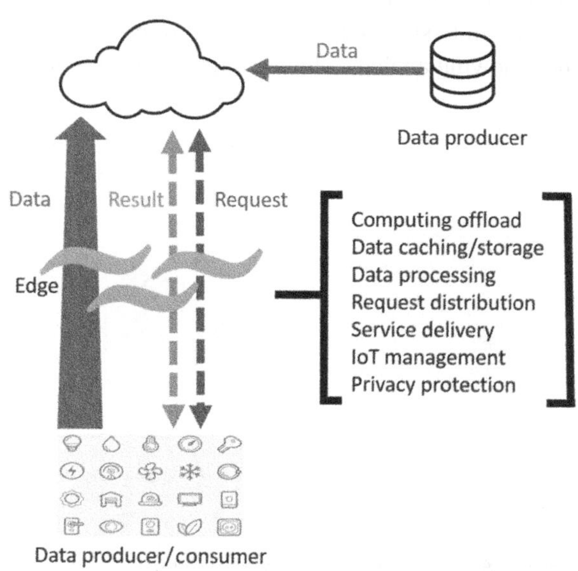

Figure 1.5 Edge computing paradigm.

"edge" can be any computing and network resources along the path between data sources and cloud data centers. The rationale of edge computing is that **computing should happen at the proximity of data source, close to the users**. We can interpret the word "close" in two ways. First, the edge computing resource and the end users may be close in the communication network. Therefore, the small network size makes it more feasible to deal with network instability (e.g., bandwidth, delay, and jitter). Second, the resource and users may be close in spatial distance, which means they share similar environmental information. The computing resources may leverage the shared information to provide personalized services and improve the user experience. Network distance and spatial distance are not correlated to each other, and it may depend on the concrete scenarios to decide which type of close (or both) is appropriate.

If we view resources on the path between IoT services and cloud services as a continuum, *edge* can be any computing, storage, and network resources on this path. Depending on the specific requirements and concrete scenarios, the edge can be one or multiple resources (nodes), as shown in Figure 1.6. It can be a smartphone or a desktop serving as the edge between body things and the cloud, a gateway in a smart home as the edge between home things and the cloud. It can also be as small as embedded devices such as wearable sensors and security cameras, or as big as a micro data center. There are a huge amount of edge resources; they are scarcely distributed around end users, independent of each other. Edge computing is dedicated to unifying these resources that are close to end users in either the communication network or spatial distance and provides computing, storage, and network services for applications.

If we understand edge computing from a biological perspective, we can make the following analogy: cloud computing is akin to the human brain, while edge

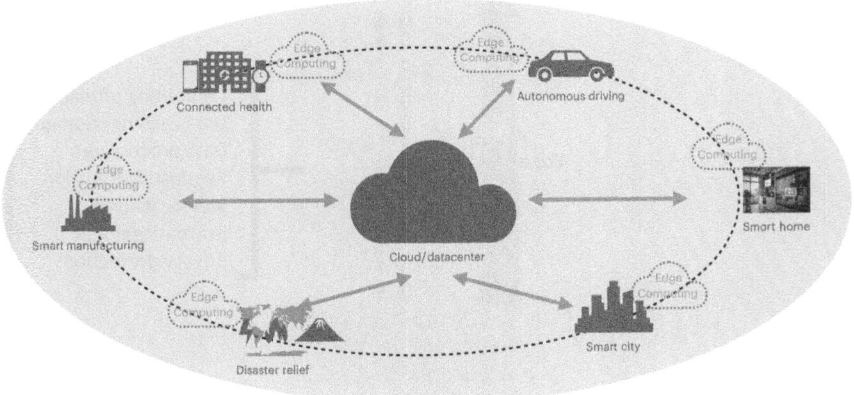

Figure 1.6 Edge computing is a continuum.

computing is akin to nerve endings. When a needle pricks the hand, a person instinctively pulls their hand back before the brain even realizes the prick because the reflex action is processed by the nerve endings. This reflex action speeds up the response, preventing further harm, while allowing the brain to focus on more complex tasks. In the future era of the IoE, it is impractical for cloud computing to act as the "brain" for every device. Instead, edge computing allows edge devices to have their own "brains."

To have an overview of the development of edge computing, we investigate the time period when it got attention and became popular. If you search for "edge computing" in Google Trends, set the *time range* to be *01/01/2010–now*, you will see a trend like Figure 1.7. "Interest over time" shows search interest relative to the highest point on the chart for the given region (in this example, we choose *Worldwide*) and time (from 2010 till 2024). A value of 100 means the term is in the peak popularity. A value of 50 means that the term is half as popular. A score of 0 means there was not enough data for this term. We can see that the attention to edge computing has been continuously increasing since 2016, reflecting its importance in the development of technology. In the years 2022 to 2023, there is a drop in "interest over time" for edge computing. At the same, another term, **Edge AI**, has been raising attention and has been getting more and more interest since 2023, when the large language model (LLM) started to dominate the artificial intelligence (AI) and machine learning (ML) market. The trend of these two words perfectly shows the focus of the research area in edge computing. Before LLM showed up, the research focus in this field was on the computing paradigm itself, such as the architectures and components of edge computing (Chapter 3), edge computing hardware and software (Sections 4.1 and 4.2), and challenges and solutions in edge computing (Chapter 5). With the development of AI and ML, especially LLMs, the world has witnessed the power of AI models. However, to make these powerful models accessible to the masses, the computing and storage resource constraints became a significant bottleneck. Therefore, the research

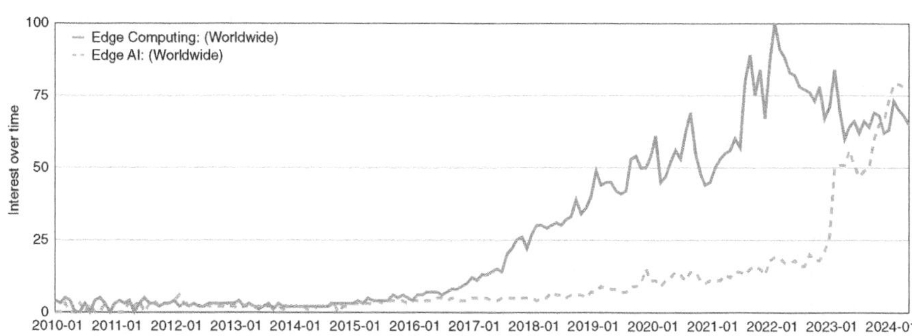

Figure 1.7 "Edge computing" and "edge AI" trends.

focus in this field has shifted to enabling edge-based AI by tackling the problem of resource constraints (Sections 4.3 and 4.4).

1.4 Summary and Practice

1.4.1 Summary

This chapter provides the definition and core concept of edge computing, emphasizing that edge computing is a continuum. The "edge" in edge computing refers to any computing, storage, and network resources along the path from the data source to the cloud computing center. The discussion of the development and challenges of big data processing and the Internet of Everything helps to understand the background of the emergence of this computing paradigm. This chapter also reviews the historical development of data-oriented computing models such as distributed databases, P2P, CDN, MEC, fog computing, and Cloud-Sea computing. Additionally, it introduces the current status of edge computing and its close connection with edge intelligence.

1.4.2 Practice Questions

1) What is the "edge"?
2) What are the main characteristics that distinguish edge computing from traditional cloud computing?
3) Identify and explain the challenges that traditional cloud computing faces in handling the large volumes of data generated by IoE devices.
4) Why is edge computing necessary in the era of the Internet of Everything?
5) Explain examples of real-world applications for each use case presented in Figure 1.6.
6) Based on the background and evolutionary history of edge computing discussed in this chapter, what do you think are the key challenges that could arise during the development of edge computing?

1.4.3 Course Projects

1) Analyze real-world case studies where edge computing is used to solve specific problems and understand why edge computing was necessary in each case. Case studies can be found from sources like https://lfedge.org/.
2) Conduct a comprehensive study of various open-source edge platforms to understand the capabilities, strengths, weaknesses, and potential use cases of each platform. The platforms to be researched could include, but are not

limited to: LF Edge Projects (EdgeX Foundry, Akraino, and Open Horizon), Kubernetes for edge (KubeEdge, K3s, and MicroK8s), OpenFaaS.

3) Explore a basic edge computing use case and present a simple prototype. For example, smart home, healthcare monitoring. The prototype will involve the design of data processing system architecture, and operate on an edge device.
4) Build a smart home environment that leverages edge computing to manage and control devices such as lights, temperature sensors, security cameras locally, instead of relying solely on cloud services.

Chapter 1 Suggested Papers

1 Mahadev Satyanarayanan. "The emergence of edge computing". In: *Computer* 50. 1 (2017), pp. 30–39.

2 Weisong Shi et al. "Edge computing: Vision and challenges". In: *IEEE Internet of Things Journal* 3. 5 (2016), pp. 637–646.

3 Weisong Shi, George Pallis, and Zhiwei Xu. "Edge computing [scanning the issue]". In: *Proceedings of the IEEE* 107. 8 (2019), pp. 1474–1481.

References

1 Akamai. *Amazon CloudFront*. https://www.akamai.com/solutions/content-delivery-network. Accessed: 2024-07-24.

2 Amazon AWS. *Amazon CloudFront*. https://aws.amazon.com/cloudfront/. Accessed: 2024-07-24.

3 Kevin Ashton. "That 'Internet of Things' thing". In: *RFID Journal* 22. 7 (2009), pp. 97–114.

4 Automotive Edge Computing Consortium. *Automotive Edge Computing Consortium*. https://aecc.org/. Accessed: 2024-07-31.

5 AWS. *IoT Edge*. https://aws.amazon.com/greengrass/. Accessed: 2024-08-26.

6 Bio-IT World. *bio-IT World Conference Edge Track*. https://kubeedge.io/. Accessed: 2024-07-31.

7 Flavio Bonomi et al. "Fog computing and its role in the Internet of Things". In: *Proceedings of the 1st Edition of the MCC Workshop on Mobile Cloud Computing*. 2012, pp. 13–16.

8 Weisong Shi Mung Chiang. *NSF Workshop Report on Grand Challenges in Edge Computing*. https://www.weisongshi.org/papers/shi16-nsfreport.pdf. Accessed: 2024-07-30.

9 Cisco. *Edge Computing Solutions*. https://www.cisco.com/c/en/us/solutions/service-provider/edge-computing.html. Accessed: 2024-05-11.

10 Cisco. *Redefine Connectivity by Building a Network to Support the Internet of Things.* https://www.cisco.com/c/en/us/solutions/service-provider/a-network-to-support-iot.html. Accessed: 2024-05-25.

11 Laura DeNardis. *The Internet in everything.* Yale University Press, 2020.

12 Edge Computing Consortium. *Introduction of Edge Computing Consortium.* http://en.ecconsortium.net/Uploads/file/20180328/1522232376480704.pdf. Accessed: 2024-07-31.

13 Exploding Topics. *Amount of Data Created Daily (2024).* https://explodingtopics.com/blog/data-generated-per-day. Accessed: 2024-05-14.

14 Iggy Fernandez. *No! to SQL and No! to NoSQL.* https://iggyfernandez.wordpress.com/2013/07/28/no-to-sql-and-no-to-nosql/. Accessed: 2024-07-24.

15 Google Cloud. *Google Distributed Cloud.* https://cloud.google.com/distributed-cloud/edge/latest/docs. Accessed: 2024-08-26.

16 Katarina Grolinger et al. "Data management in cloud environments: NoSQL and NewSQL data stores". In: *Journal of Cloud Computing: Advances, Systems and Applications* 2 (2013), pp. 1–24.

17 Adam Grzywaczewski. *Training AI for Self-Driving Vehicles: the Challenge of Scale.* https://developer.nvidia.com/blog/training-self-driving-vehicles-challenge-scale/. Accessed: 2024-08-26.

18 Jayavardhana Gubbi et al. "Internet of Things (IoT): A vision, architectural elements, and future directions". In: *Future Generation Computer Systems* 29. 7 (2013), pp. 1645–1660.

19 Yun Chao Hu et al. "Mobile edge computing—A key technology towards 5G". In: *ETSI White Paper* 11. 11 (2015), pp. 1–16.

20 IBM. *What is edge computing?* https://www.ibm.com/topics/edge-computing. Accessed: 2024-05-11.

21 Insurance Information Institute. *Background on: Self-driving cars and insurance.* https://www.iii.org/article/background-on-self-driving-cars-and-insurance. Accessed: 2024-08-26.

22 Hosagrahar V Jagadish et al. "Timber: A native XML database". In: *The VLDB Journal* 11 (2002), pp. 274–291.

23 KPMG. *Corporate data responsibility: Bridging the consumer trust gap.* https://kpmg.com/us/en/articles/2023/bridging-the-trust-chasm.html. Accessed: 2024-08-26.

24 KubeEdge Project Authors. *KubeEdge.* https://kubeedge.io/. Accessed: 2024-07-31.

25 Microsoft Azure. *Azure Stack.* https://azure.microsoft.com/en-us/products/azure-stack/. Accessed: 2024-08-26.

26 Microsoft Azure. *IoT Edge.* https://azure.microsoft.com/en-us/products/iot-edge/. Accessed: 2024-08-26.

27 Dejan S Milojicic. *Peer-to-peer computing.* 2002.

28 OpenFog Consortium. *OpenFog Reference Architecture for Fog Computing.* https://www.iiconsortium.org/pdf/OpenFog_Reference:Architecture_2_09_17.pdf. Accessed: 2024-07-31.

29 Gang Peng. “CDN: Content distribution network”. In: *arXiv preprint cs/0411069* (2004).

30 Michael Rabinovich, Zhen Xiao, and Amit Aggarwal. “Computing on the edge: A platform for replicating internet applications”. In: *Web Content Caching and Distribution: Proceedings of the 8th International Workshop.* Springer. 2004, pp. 57–77.

31 Jayashree Ravi, Weisong Shi, and Cheng-Zhong Xu. “Personalized email management at network edges”. In: *IEEE Internet Computing* 9. 2 (2005), pp. 54–60.

32 Mahadev Satyanarayanan. “The emergence of edge computing”. In: *Computer* 50. 1 (2017), pp. 30–39.

33 Mahadev Satyanarayanan et al. “The case for VM-based cloudlets in mobile computing”. In: *IEEE Pervasive Computing* 8. 4 (2009), pp. 14–23.

34 Weisong Shi et al. “Edge computing: Vision and challenges”. In: *IEEE Internet of Things Journal* 3. 5 (2016), pp. 637–646.

35 Statista. *Number of Internet of Things (IoT) connected devices worldwide from 2019 to 2023, with forecasts from 2022 to 2030.* https://www.statista.com/statistics/1183457/iot-connected-devices-worldwide/. Accessed: 2024-05-25.

36 Harald Sundmaeker et al. “Vision and challenges for realising the Internet of Things”. In: *Cluster of European Research Projects on the Internet of Things, European Commission* 3. 3 (2010), pp. 34–36.

37 U.S. National Science Foundation. *NSF 16-586: NSF/Intel Partnership on Information-Centric Networking in Wireless Edge Networks (ICN-WEN).* https://new.nsf.gov/funding/opportunities/nsfintel-partnership-information-centric/505310/nsf16-586/solicitation. Accessed: 2024-07-30.

38 UN Trade and Development. *Digital economy report 2024.* https://unctad.org/publication/digital-economy-report-2024. Accessed: 2024-08-26.

39 Luis M Vaquero and Luis Rodero-Merino. “Finding your way in the fog: Towards a comprehensive definition of fog computing”. In: *ACM SIGCOMM Computer Communication Review* 44. 5 (2014), pp. 27–32.

40 Zhi-Wei Xu. “Cloud-sea computing systems: Towards thousand-fold improvement in performance per watt for the coming zettabyte era”. In: *Journal of Computer Science and Technology* 29. 2 (2014), pp. 177–181.

41 Ashkan Yousefpour et al. “All one needs to know about fog computing and related edge computing paradigms: A complete survey”. In: *Journal of Systems Architecture* 98 (2019), pp. 289–330.

2

Fundamentals of Edge Computing

In the context of the industrialization of emerging information technologies, no new technology arises in a vacuum. Each new development is a response to the growing demands of new applications for high performance, real-time processing, low energy consumption, and low latency, which highlight the limitations of existing information systems in computation, storage, and transmission. In the era of the Internet of Everything (IoE) and big data, the volume of data generated by network edge devices has increased dramatically. Traditional cloud-based big data-processing technologies are gradually unable to meet the real-time processing and low energy consumption demands of users. It is within this context that edge computing has emerged, addressing the shortcomings of current big data-processing technologies and rapidly gaining significant attention from both industry and academia over the past decade.

2.1 Distributed Computing

Distributed computing [6, 29] involves connecting numerous computer nodes via the Internet to break down a computational task, which a single computer cannot handle, into multiple tasks that are distributed among various computers in the network. The edge node executes its assigned task, and the results are integrated into the final output, which is then returned. This process represents computation executed on a distributed system. Distributed computing relies on multiple distributed computing units interconnected by high-speed networks to perform high-performance computations. It is characterized by meeting user demand and resource availability, enabling resource sharing among different nodes. The main challenges it faces include heterogeneity, scalability, faculty tolerance, and concurrency.

Edge Computing: Systems and Applications, First Edition. Lanyu Xu and Weisong Shi.

Companion Website: www.wiley.com/go/xu/computing

2.1.1 Distributed Computing Technologies

Distributed computing technologies primarily include middleware technology [8], grid computing technology [5], mobile agent technology [33], P2P technology [22], and web service technology [7].

- **Middleware Technology**: Middleware is situated between the operating system and distributed application software, used to mask the heterogeneity of operating systems and network protocols in a distributed environment. IBM developed a customer information control system (CICS) with middleware functions in the 1960s. Early middleware software had relatively simple functions, primarily providing message communication and transaction processing. As the demand for middleware applications diversified, middleware technology evolved into several categories: remote procedure call-based middleware, message-oriented middleware, database middleware, and object-oriented middleware. Among these, object-oriented middleware has become the mainstream technology for middleware platforms.
- **Grid Computing Technology**: Grid computing integrated geographically dispersed hardware and software resource through high-speed networks to complete large-scale complex computation and data-processing tasks. It generally refers to two widely used subtypes in distributed computing: one is online computation or storage provided as a service supported by distributed computing resources; the other is a virtual supercomputer formed by loosely connected computer networks to execute large-scale computation tasks. In terms of grid architecture, grid computing is mainly divided into two types: one is the five-layer hourglass structure represented by the Globus project, and the other is the open-grid services architecture (OGSA) that integrates with web services. The primary difference between the two structures is that the former is protocol-centric, while the latter is service-centric.
- **Mobile Agent Technology**: Mobile agents can autonomously and automatically migrate within heterogeneous networks and distributed computing environments, and communicate with other agents. In different network structures, mobile agents follow certain principles to locate matching resource information, perform tasks on behalf of clients, and autonomously generate subagents as ended. In a mobile agent system, each agent works independently and can collaborate to complete tasks when necessary.
- **P2P Technology**: By linking terminal devices in the network and integrating idle resources, P2P technology maximizes resource sharing and distributed computing. In a P2P network, each node contributes its idle resources and uses resource discovery mechanisms to find available resources on other nodes, enabling resource sharing. P2P technology can maximize the utilization of network resources. However, due to its characteristics of openness,

self-organization, autonomy, and distribution, network users can dynamically and anonymously join or leave the system. Consequently, P2P technology may face practical issues such as copyright infringement, lack of management mechanisms, network pollution, and malicious attacks.

- **Web service Technology**: Web service technology is an extension of object/component technology on the Internet and a type of distributed computing technology deployed on the web. The primary goal of web service technology is to construct a platform- and language-independent general technical layer on top of existing heterogeneous platforms, allowing applications on different platforms to run smoothly. This technology addresses the issue of limited interoperability, thereby improving and expanding the functionality of distributed computing.

2.1.2 Distributed System Platforms

The numerous challenges of big data environments have spurred the use of distributed computing technology and the growth of distributed systems. Today, Hadoop [32], Spark [36], and Storm [4] are among the most widely used platforms.

Hadoop, developed by the Apache Software Foundation, is a distributed computing framework whose core components include a distributed file system (Hadoop distributed file system, HDFS [27]), a programming model (MapReduce [13]), and a distributed structured data table (HBase [17]). These correspond to the open-source implementations of Google's core cloud computing technologies: GFS [18], MapReduce, and Bigtable [11]. MapReduce abstracts the parallel computing process on large-scale distributed systems into two functions: Map and Reduce.

When a task is submitted to the Hadoop platform, it is divided into multiple chunks. The JobTracker assigns the currently idle TaskTracker to perform parallel Map operations on these chunks. The RecordReader generates <K, V> key–value pairs from the split data chunks for parallel execution across different nodes. The intermediate records produced by the Map tasks are further divided into multiple chunks, with the JobTracker again assigning idle TaskTrackers to perform parallel Reduce operations on these chunks. The final results are written to output files and are managed by HDFS.

However, because MapReduce stores intermediate results on disk during distributed computing, especially during iterative computations in data mining where previous results need to be frequently accessed and used, the system's performance is significantly affected.

To address the performance degradation caused by Hadoop MapReduce frequent reading and writing to the file system, the AMP lab at the University of California, Berkeley developed Spark, a memory-based computing platform.

The key technology behind Spark is the creation of Resilient Distributed Datasets (RDDs), which support data being stored in memory, thereby enabling an in-memory MapReduce architecture. By using RDDs, the MapReduce process avoids writing intermediate results back to the HDFS file system, significantly boosting computational efficiency. This improvement is especially pronounced in interactive computations, where Spark can be over 100 times faster than Hadoop. However, while Spark alleviates the problem of frequent disk I/O, it is highly memory-intensive.

Apache Storm, an open-source distributed real-time computation system developed by Twitter (now known as X), is designed to efficiently process massive streams of data. Storm supports multiple programming languages and provides a set of primitives for real-time computation, greatly reducing the development cycle for large-scale real-time processing.

Storm is widely used in various applications such as online machine learning, real-time analytics, distributed remote procedure calls (RPCs), continuous computation, and Extract, Transform, Load (ETL). Real-time applications requiring Storm are packaged into task topology and submitted for execution. Once submitted, these task topology will continue to run until explicitly terminated. A task topology is composed of a series of spouts and bolts arranged in a directed acyclic graph (DAG). Spouts are responsible for reading data from external sources, while bolts process the data received from spouts or other bolts. By cascading spouts and bolts, the system completes the desired computation tasks.

In summary, distributed computing has evolved from the MapReduce architecture implemented on the Hadoop platform to distributed systems like Storm that support stream processing. This evolution aims to meet the increasing real-time demands of big data processing.

2.2 The Basic Concept and Key Characteristics of Edge Computing

Having established a foundational understanding of distributed computing technologies and platforms, we can now shift our focus to a more specialized domain within this field: edge computing. This transition is essential as edge computing builds upon the principles of distributed computing but applies them in the context of optimizing data processing closer to the data source. This section will delve into the basic concepts and key characteristics of edge computing, highlighting how it differs from and extends the ideas we explored in distributed computing.

2.2.1 The Basic Concept

In the era of the IoE, connectivity extends beyond just things in the IoT. It includes interactions between people and things, featuring contextual awareness, enhanced computing capabilities, and advanced sensing abilities. In this interconnected model, people and information are integrated into the Internet, creating a network with billions or even trillions of connected nodes. IoE is based on physical networks, combining network intelligence, collaboration among all connected things, and visualization functions.

Sensors, smartphones, wearable devices, and smart appliances will all become part of the IoE, generating massive amounts of data. However, the current cloud computing model lacks the network bandwidth and computing resources to efficiently handle this data [1, 20]. Figure 2.1 illustrates the traditional cloud computing model. In this model, source data is sent from producers to the cloud, and data consumers, such as smartphones, personal computers, and even autonomous driving cars, send usage requests to the cloud center. In the figure, solid lines represent source data being sent by data producers to the cloud center, dashed lines represent data consumers sending usage requests to the cloud center, and dotted lines show the cloud center sending results back to the data consumers.

Cloud computing uses extensive cloud resources to process data, but in the IoE environment, the traditional cloud computing model cannot effectively meet application needs. The main reasons are: (1) sending massive amounts of data from edge devices directly to the cloud leads to network bandwidth overload and wasted computing resources; (2) privacy protection issues in the traditional cloud computing model pose significant challenges in the IoE architecture; (3) most edge device nodes in the IoE architecture have limited energy, while wireless transmission modules like global system for mobile communications (GSM) and Wi-Fi consume a lot of power.

To address these issues, leveraging the existing computing capabilities of edge devices by migrating all or part of the application service tasks from the cloud center to the edge devices will help reduce energy consumption [30].

Currently, edge devices not only consume data, such as when users watch online videos on smartphones, but also produce data, like when people share

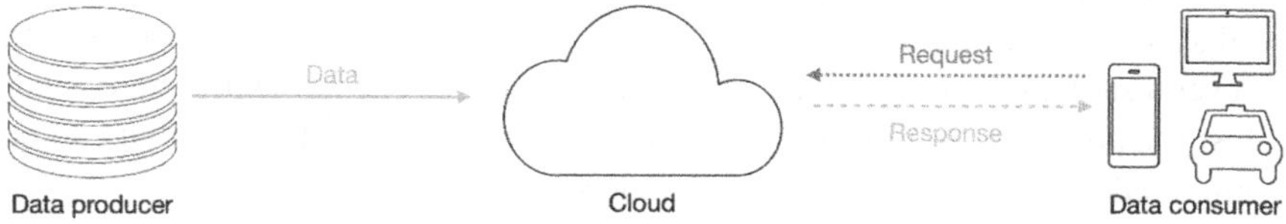

Figure 2.1 Traditional cloud computing model.

photos and videos on platforms like Facebook and X. This shift from being data consumers to data producers requires edge devices to have more powerful computing capabilities. For example, every single minute, X users send 360 K tweets, Instagram users send 694 K reels through direct messages per minute, ChatGPT users send 6944 prompts, and viewers watch 43 years' worth of streaming content [15]. Additionally, autonomous driving vehicles generate vast amounts of sensor and video data every second, posing significant challenges to bandwidth and computing resources for processing and uploading the generated data. These large volumes of images, videos, and sensor data require significant bandwidth when uploaded to cloud computing centers. To address this, preprocessing can be performed on edge devices before uploading the source data to the cloud, reducing the amount of data transmitted and alleviating bandwidth load. Moreover, processing personal health data and other sensitive information on edge devices enhances privacy protection for users [16, 28].

Edge computing is a new computing model that executes computations at the network's edge. It involves two types of services: downstream cloud services and upstream IoE services. In this model, "the edge" encompasses all computing, storage, and network resources along the path from the data source to the cloud computing center. As illustrated in Figure 1.5, edge computing relies on a bidirectional computational flow. Cloud centers collect data not only from databases but also from edge devices like sensors and smartphones. These devices function as both data producers and consumers, making the data exchange between the endpoint devices and the cloud center bidirectional. Edge devices do more than request content and services from the cloud; they also undertake various computation tasks, including data storage, processing, caching, device management, and privacy protection. Enhancing the design of the hardware platforms and key software technologies of edge devices is crucial to meet the demands for reliability and data security in edge computing.

From a functional standpoint, the edge computing model is a distributed computing system characterized by elastic management, collaborative execution, environmental heterogeneity, and real-time processing capabilities. It shares similarities with streaming computation models but also includes unique features that are outlined below (Figure 2.2):

- **Divisibility of applications/services**: Applications or services suitable for edge computing must be divisible. This means a task can be decomposed into several subtasks, which can be executed at the edge. The key aspect here is not just the ability to split tasks but also their migratability. Only tasks that can be migrated for edge processing meet the necessary criteria.
- **Distributability of data**: This feature defines edge computing and dictates the requirements for data sets being processed. If the data lacks distributability, then

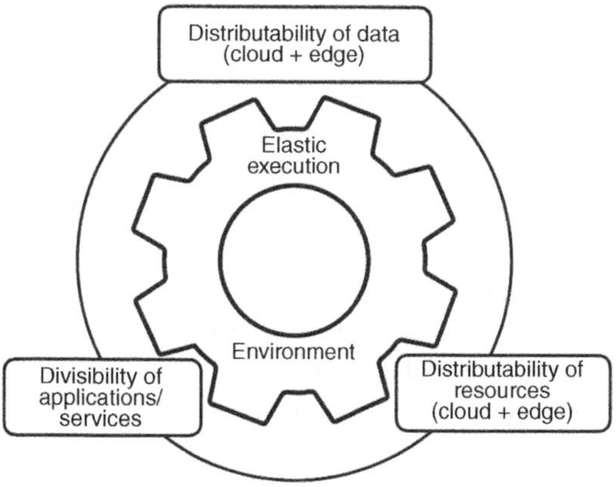

Figure 2.2 Edge computing model characteristics.

the model resembles a centralized cloud computing framework. Distributability needs to address data from various sources, typically generated by producers creating large volumes of data.

- **Distributability of resources**: As the data in edge computing models is inherently distributed, so too must be the computing, storage, and communication resources required to process this data. An edge system can only adhere to the true model of edge computing if it has the resources necessary for processing and computing data at the edge.

2.2.2 The Key Characteristics

2.2.2.1 Compute Migration

In traditional cloud computing models, compute migration strategies typically involve shifting compute-intensive tasks to well-resourced data centers. However, in the context of the IoE, the vast data volumes generated by numerous edge devices cannot be efficiently transmitted to these centers due to limited bandwidth. Even though cloud centers have significantly lower computational latency compared to edge devices, the substantial data transmission overhead can hinder overall system performance.

Thus, the compute migration strategy in edge computing should focus on minimizing the amount of data that needs to be transmitted across the network, rather than relocating compute-intensive tasks to edge devices.

The edge computing strategy includes conducting partial or complete preprocessing of data directly at the network edge, where data is initially collected or generated by the edge devices. This preprocessing aims to filter out unnecessary data, reducing the bandwidth demand. Moreover, it is crucial to dynamically allocate tasks based on the current computational load of the edge devices to avoid overloading any single device, which could degrade system performance.

Key considerations in compute migration include determining which tasks are suitable for migration, deciding on a migration strategy, selecting specific tasks for migration, and deciding whether to perform partial or complete migrations. The decisions regarding compute migration should be tailored to the application model, considering whether the application can be migrated, whether the required data volume for processing is accurately known, and whether the tasks can be efficiently synchronized postmigration.

Ultimately, compute migration technology should strive to find the optimal balance between energy consumption, computational latency at the edge, and the volume of data transmitted, thus enhancing the performance and efficiency of the edge computing model.

2.2.2.2 5G and 6G Communication Technologies

The fifth and sixth generations (5G and 6G) of mobile telecommunications systems aim to deliver higher data speeds, ultralow latency, more reliability, massive network capacity, increased availability, and a more uniform user experience to more users.

The 5G technology standard for cellular networks, which cellular phone companies began developing worldwide in 2019, is the planned successor to the 4G networks, which provide connectivity to most current cell phones. 5G networks are predicted to have more than 1.7 billion subscribers worldwide by 2025, according to the GSM Association [31]. Compared to 4G, 5G significantly increases the speed and responsiveness of wireless networks, supports far more devices at high data rates, and reduces latency, which is beneficial for new technologies such as autonomous driving, virtual reality, and the IoT.

Although still in the early stages of development and standardization, 6G networks are expected to succeed in 5G. Predictions suggest that 6G will enable even higher speeds and lower latency, with the integration of advanced technologies like artificial intelligence (AI) and sophisticated satellite networks. It is anticipated to support even more innovative applications, potentially including advanced augmented reality, high-fidelity mobile holograms, and greater integration of physical and digital realities.

To meet the diverse application scenarios and business demands, 5G and 6G networks will require a universal, scalable, and easily extendable network architecture. This will also involve integrating advanced technologies such as software-defined networks (SDNs) and network functions virtualization (NFV).

5G and 6G technologies are pivotal in the edge computing model. Edge devices, tasked with processing either part or all of the computational duties and filtering out redundant and sensitive information, still need to upload intermediate or final data to cloud centers. Thus, 5G and 6G technologies are critical in reducing data transmission delays for mobile edge devices, ensuring faster and more efficient communication.

2.2.2.3 Advanced Storage Systems

As computer processors continue to advance rapidly, the speed disparity between storage systems and processors has become a significant bottleneck in overall system performance. Edge computing requires robust real-time capabilities for data storage and processing. Compared to traditional embedded storage systems, edge-computing storage solutions offer lower latency, increased capacity, and enhanced reliability. These systems must handle data characterized by high immediacy, diversity, and interconnectivity, ensuring continuous storage and preprocessing of edge data. Therefore, efficiently managing and accessing continuous, real-time data is a critical focus in the design of storage systems for edge computing.

Presently, nonvolatile memory (NVM) is extensively used in embedded systems and large-scale data processing. Storage devices based on NVM, such as NAND Flash, PCRAM, and RRAM, provide significantly better read and write performance than traditional mechanical hard drives [25], thus effectively mitigating the I/O limitations of existing storage systems. However, most traditional storage system software stacks, designed primarily for mechanical hard drives, do not fully leverage the maximum capabilities of NVM.

As edge computing rapidly advances, NVM, characterized by high density, low energy consumption, low latency, and high read/write speeds, is increasingly being deployed in edge devices. However, the integration of NVM within edge systems faces several challenges:

- **Rapid Technological Development vs. Software Support**: The fast-paced advancement of NVM technologies is not mirrored by the development of supporting software stacks, leading to a "software bottleneck." This mismatch hampers the ability of storage systems to fully utilize the speed and efficiency of modern NVM technologies.
- **Diverse Application Requirements**: Edge computing demands a variety of applications for emerging storage architectures. A crucial area of research involves maximizing the performance, energy efficiency, and capacity benefits of nonvolatile storage systems. Key issues include optimizing nonvolatile memory for timely edge data processing and simplifying storage system management in complex edge environments.
- **Reliability in Challenging Environments**: Edge environments require robust read/write capabilities and high data reliability. Ensuring data reliability in NVM under complex external conditions and resource constraints is a pivotal

concern. Factors affecting data reliability include consistency issues in NVM, targeted malicious wear attacks, and the lifespan and failure rates of the storage media.

Addressing these challenges is essential for optimizing the use of NVM in edge computing, necessitating focused research and development in both hardware and software domains.

2.2.2.4 Lightweight Libraries and Kernels

Unlike large servers, edge devices are constrained by their hardware capacity and often cannot support the operation of heavy software applications. For example, while advanced RISC machines (ARM) processors continue to increase in speed and decrease in power consumption, they still lack the capability to handle complex data-processing applications effectively. For instance, running Apache Spark optimally requires at least an 8-core CPU and 8 GB of memory. In contrast, the lightweight library Apache Quarks [3] can only perform basic data-processing tasks and is unsuitable for advanced analytics.

Moreover, the network edge is populated with a diverse array of devices from various manufacturers, characterized by significant heterogeneity and varying performance levels, making application deployment on these devices a complex challenge. Virtualization technology often serves as a solution; however, traditional virtual machine (VM)-based virtualization is too resource-intensive and slow for edge environments, where swift responses are crucial.

Instead, edge competing models should adopt lightweight virtualization technologies that align with the limited resources of edge devices. Lightweight libraries and kernels are particularly valuable in this context, as they consume fewer resources and time, thereby optimizing performances. Docker is an example of such a technology that utilizes containerization. Docker containers are much more resource-efficient compared to VMs because they virtualize at the operating system level and share the host system's kernel, rather than requiring a full operating system for each instance. This allows Docker to provide isolated environments for applications using minimal resources, making it ideal for deploying applications on resource-constrained edge devices. Docker not only enhances the scalability and deployment speed but also maintains consistent environments across development, testing, and production, reducing compatibility issues. Therefore, Docker and similar container-based technologies are indispensable for optimizing performance and resource utilization in edge computing.

2.2.2.5 The Edge Computing Programming Model

In the cloud computing model, users write applications and deploy them to the cloud, where cloud service providers maintain the servers. This model allows users to remain largely unaware of the application's operation, benefiting from the infrastructure's transparency. Typically, user programs are written and compiled on the target platform and run on cloud servers.

In contrast, the edge computing model involves migrating some or all computing tasks from the cloud to edge nodes. These nodes often exist on heterogeneous platforms, each with different operating environments, presenting significant deployment challenges for programmers. The traditional programming models are inadequate for the setting, highlighting the need for new programming models tailored to edge computing.

To address this, we propose a concept known as a "computation stream." This concept represents the data transmission path and the sequence of computations or functions performed on the data. These functions, integral to an application, occur along data paths that enable computational applied at the source device, edge nodes, and cloud environments to enable efficient distributed data processing.

The evolution of the programming model necessitates new runtime libraries, which are essential for implementing language functions and providing runtime support. This foundation is critical in edge computing, where programming model changes demand novel runtime libraries. These libraries should offer specific application interfaces that simplify application development for programmers, thus ensuring that applications can adapt to the diverse and dynamic nature of edge environments.

Building on this foundation, we introduce the "Firework Model" [37], a novel programming model designed specifically for edge computing. This model includes two components: the firework model manager and the firework model nodes. The manager decomposes service requests into several subtasks and distributes them among participants, where each subtask is executed locally on the participant's device. The nodes provide end-users with a suite of predefined functional interfaces, facilitating easier access and interaction. In this model, all nodes must register their datasets and functionalities, which are abstracted into a data view. These registered data views are visible to all participants within the same model, allowing any participant to combine these views for specific data analyses tailored to particular scenarios. This approach not only leverages the capability of new runtime libraries but also aligns with the need for flexible and efficient data processing at the edge, embodying the principles of software-defined computations in practical, deployable models.

2.3 Edge Computing vs. Cloud Computing

With a clear grasp of edge computing's fundamental concepts and its distinct characteristics, it is crucial to understand how this technology fits into the broader landscape of computing. Specifically, we will explore the relationship between edge computing and cloud computing. This relationship is pivotal for addressing big data challenges and leveraging the strengths of both paradigms. This section will analyze how edge computing complements cloud computing, discuss their interaction with big data, and evaluate the advantages of edge computing and the challenges it faces.

2.3.1 The Concept of Cloud Computing

Cloud computing [10, 14, 19] is a service delivery model that provides scalable distributed computing capabilities by accessing computational, network, and storage resources in data centers over a network. This model leverages existing resources and uses virtualization technology [9, 26] to create a shared resource pool of many computers. It offers powerful computational and management capabilities and can dynamically partition and allocate resources to meet diverse user needs, ensuring efficient service delivery.

Cloud computing is an evolution of parallel computing, distributed computing, and grid computing, or essentially, a commercial realization of these computational science concepts. Cloud computing is generally divided into three service types: Infrastructure as a Service (IaaS), Platform as a Service (PaaS), and Software as a Service (SaaS). As cloud computing continues to advance, different cloud computing solutions are increasingly converging and integrating with one another.

From the current research status, cloud computing has the following characteristics:

- **Large-Scale Cloud Server**: Major IT companies like Google and Microsoft have cloud computing platforms with hundreds of thousands of servers. Even private cloud projects for typical IT enterprises may involve hundreds or thousands of servers. This vast scale provides users with substantial computing power and massive storage capacity.
- **High Reliability**: Cloud computing platforms are designed with distributed server clusters, making single-pint failures inevitable. To ensure high reliability, cloud computing centers employ fault-tolerance mechanisms such as replication strategies and homogeneous interchange of computing nodes.
- **Scalability**: Cloud computing dynamically allocates or releases resources based on specific user needs. When demand increases, cloud computing can quickly provide matching resources, offering high-speed and flexible scalability. Similarly, resources can be released when they are no longer needed, thanks to the inherent scalability of cloud computing.
- **Virtualization**: Cloud computing integrates resources dispersed across different geographical locations into a logically unified shared resource pool

through virtualization technology. Users can request services from the cloud computing center anytime and anywhere via the Internet. Virtualization hides the heterogeneity of underlying device resources, enabling unified scheduling and deployment of all resources. The cloud infrastructure is transparent, and users do not need to worry about the specific location of these resources. Therefore, virtualization is not only the foundation of cloud computing but also a defining characteristic.

2.3.2 The Big Data Era

In the era of the IoE, the large-scale application and widespread development of new technologies such as mobile devices and the Internet have brought data informatization into the big data era. In May 2011, McKinsey & Company first introduced the concept of "big data" in their report "Big Data: The Next Frontier for Innovation, Competition, and Productivity" [24]. Since then, the potential value of big data has gained significant attention from national strategic research departments worldwide, elevating it to a matter a national importance. The United States launched the "Big Data Research and Development Initiative," South Korea is actively advancing its "Big Data Center Strategy," and China has formatted the "13th Five-Year Plan for Big Data Industry Development." Big data is poised to encompass all fields of economic and social development and become a new driving force for national economies.

In recent years, big data has become a major focus for academics, industry professionals, and governments worldwide. Leading journals such as Nature and Science have explored the challenges and opportunities associated with big data. As data resources grown in importance, a nation's ability to harness and effectively use this data is increasingly seen as a key factor in its overall strength and influence. Big data-processing technologies cover essential areas such as the collection, storage, cleaning, analysis, mining, visualization, and privacy protection of massive and diverse datasets. The integration of edge computing and cloud computing technologies is a primary approach to addressing significant challenges related to the storage, transmission, and processing of big data. By leveraging the strengths of both edge and cloud computing, it is possible to balance big data-processing tasks efficiently and optimize bandwidth and storage requirements.

Let's examine two examples to understand the relationship between edge computing and big data processing. With the development of the IoE, both video big data [2] and medical big data [21] have high real-time demands for storage, transmission, and computation.

Video big data is generated in high-definition formats, which currently experience high latency during transmission and processing, especially in real-time scenarios such as object detection and localization. In current video big data processing, video data is typically transmitted to a central big data center for processing. These centers, with their robust computing power, handle the intelligent computation and storage of video data. However, with the widespread adoption of high-definition (1080P) and ultra-high-definition (4K) video, and the increasing number of video surveillance devices, the volume of video data has grown significantly. Transmitting this large volume of video data to the data center for analysis puts a heavy load on network bandwidth and consumes considerable resources for data processing. Additionally, the performance of video data processing is often low due to the sheer volume of data, resulting in reduced real-time processing capabilities. This delay can directly impact the ability to make timely decisions in public safety scenarios involving emergencies.

Therefore, applying edge computing technology to video big data processing is critically important. By leveraging edge computing, a portion or all of the video processing tasks can be performed at the video surveillance terminal. This approach reduces the amount of data that needs to be transmitted and lowers processing costs, thereby enhancing the efficiency and real-time capabilities of video big data processing.

Medical big data forms the backbone of intelligent healthcare. It involves sharing and collaborating on data from various sources, including hospitals, pharmaceutical manufacturers, pharmacies, and patients. Hospitals possess vast amounts of patient records, drug demand information, and disease distribution data. Pharmaceutical manufacturers and pharmacies have drug information and patient purchase data, while patients' medical data is also highly valuable. Collectively, this is known as medical big data.

A key challenge in utilizing the value of medical big data for intelligent healthcare services is multi-edge collaborative data processing. Given the sensitive nature of this data, the critical issue is how to leverage edge computing technology to handle sensitive and private data at the edge while simultaneously sharing medical big data to realize its full value.

In addition to the aforementioned video and medical big data, there are also significant needs for edge computing in sectors like smart grids and smart manufacturing. In the context of the IoT, integrating cloud computing with edge computing models offers an effective solution for tackling challenges related to big data collaboration, processing loads, transmission bandwidth, and data privacy protection.

2.3.3 Edge Computing vs. Cloud Computing

In the context of the IoE, application services require low latency, high reliability, and data security. However, the traditional cloud computing model falls short in meeting these needs, particularly regarding real-time performance, privacy

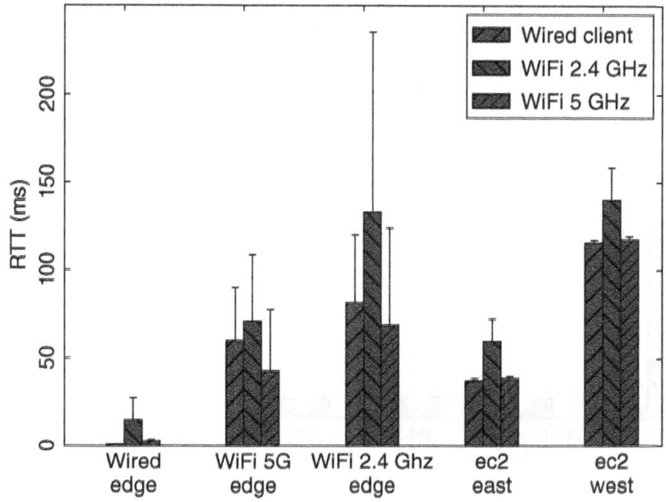

Figure 2.3 Round trip time between client and edge/cloud.

protection, and energy consumption. Edge computing models leverage the computational capabilities of edge devices, performing some or all computations and processing privacy-sensitive data at the edge. This approach reduces the computational load, transmission bandwidth, and energy consumption of cloud computing centers. The following examples illustrate the benefits of edge computing.

Yi et al. [35] tested data transmission latency and bandwidth between user nodes and either edge nodes or cloud nodes in the network, as shown in Figures 2.3 and 2.4. Amazon EC2 East and Amazon EC2 West represent two cloud nodes located in different geographic regions of the United States. Wired edge nodes, WiFi 5 GHz edge nodes, and WiFi 2.4 GHz edge nodes represent three types of edge node connections to the user's router. The tests were conducted in Washington, D.C. (near the Amazon EC2 East cloud).

The results show that when edge nodes are connected to the user's network via wired connections, the round-trip time is significantly better than that of cloud nodes. When edge nodes are connected wirelessly, the round-trip time is between the round-trip times of the two cloud nodes but is less stable due to the lower speed and stability of wireless channels compared to wired channels. Bandwidth benchmarks indicate that edge nodes connected via wired clients and WiFi 5 GHz have noticeably higher bandwidth compared to the other three types. Edge nodes using WiFi 2.4 GHz have performance levels between the two cloud nodes, primarily because WiFi 2.4 GHz bandwidth is more limited, as shown by the performance of user nodes connected via WiFi 2.4 GHz.

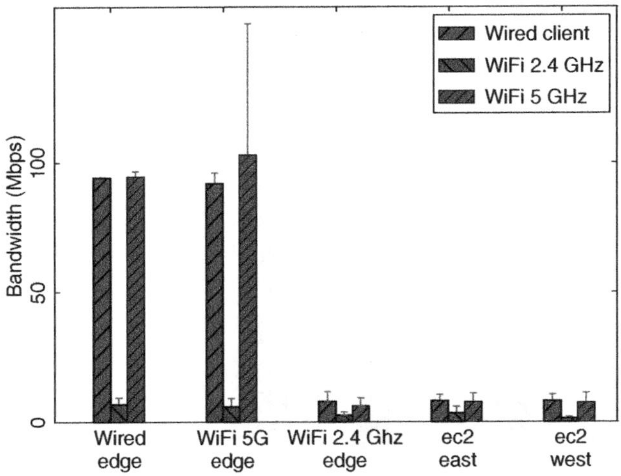

Figure 2.4 Bandwidth between client and edge/cloud.

In summary, when edge nodes have high-quality connections, their service quality surpasses that of cloud nodes. Edge nodes offer lower latency and higher bandwidth compared to cloud nodes, while cloud nodes can serve as backup computing nodes to prevent edge node saturation and handle longer request response times.

Xu et al. [34] compared the processing time of the edge and the cloud on audio command understanding. The cloud node is provided by Google Dialogflow service, the edge node is set on a Raspberry Pi. When processing the Fluent Speech Commands dataset [23], the cloud node demonstrated long latency and unstable performance in response, while the edge node solution can eliminate these drawbacks (as Figure 2.5 shows).

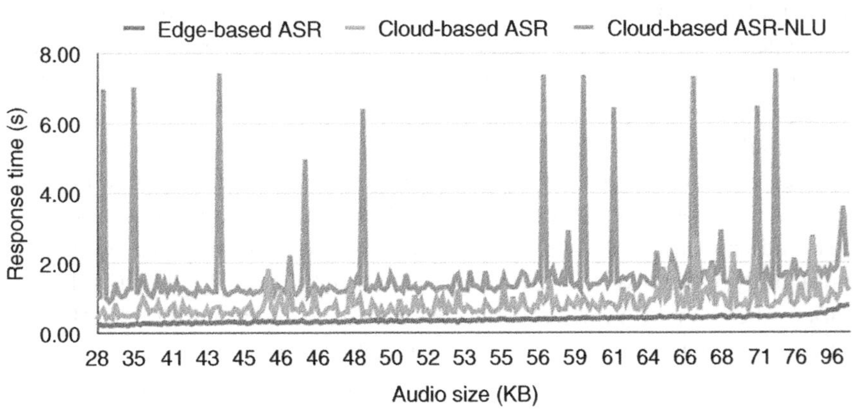

Figure 2.5 Round trip time for processing audio command on edge and cloud.

Table 2.1 Edge computing vs. cloud computing.

Comparison content	Edge computing	Cloud computing
Target applications	IoT or mobile applications	General Internet applications
Location of server nodes	Edge network (gateways, Wi-Fi access points, and cellular base stations, etc.)	Data center
Client–server communication network	Wireless local area network (WLAN), 5G/6G, etc.	Wide area network (WAN)
Number of serviceable devices (Users)	Billions	Millions
Types of service provided	Services based on local information	Services based on global information

We present the comparison between edge computing and cloud computing in Table 2.1. It is evident that edge computing is not intended to replace cloud computing but rather to complement and extend it, providing a better computing platform for mobile computing, the IoT, and other applications. The edge computing model leverages the powerful computational capabilities and vast storage of cloud computing centers while addressing the need for processing large volumes of data and private information locally on edge devices. This helps meet requirements for real-time performance, privacy protection, and reduced energy consumption. The architecture of edge computing follows a "device-edge-cloud" three-layer model, where each layer can provide resources and services for applications. This allows applications to choose the optimal configuration for their needs.

2.3.4 Advantages and Challenges of Edge Computing

The edge computing model transfers some or all computational tasks from traditional cloud data centers to locations closer to where data is generated. As noted earlier in Section 1.1, the three Vs of big data—volume, velocity, and variety—highlight the unique strengths of edge computing. In this subsection, we'll contrast centralized big data processing, typically exemplified by cloud computing models, with edge-focused big data processing, exemplified by edge computing models. We aim to highlight the advantages of edge computing from a macroscopic perspective, illustrating why it can be a more effective choice in various scenarios.

In the era of centralized big data processing, the predominant data types included text, audio/video, images, and structured databases, with volumes typically at the petabyte (PB) level. During this period, the cloud computing model did not require high real-time processing capabilities. However, in the era of edge-oriented big data processing, under the backdrop of the IoE, the types of data have become significantly more diverse and complex. Notably, sensory data from interconnected devices have surged, turning what were once mere consumer devices into active data producers. Additionally, this era is characterized by a crucial demand for real-time data processing, with data volumes now exceeding zettabytes (ZB).

In response to these changes, the need for real-time processing and the increase in data volume has necessitated migrating some computational tasks from traditional cloud centers to network edge devices. This shift aims to enhance data transmission performance, ensure timely processing, and reduce the computational load on cloud computing centers.

The unique data characteristics of the edge big data era have driven the development of edge computing models. However, the relationship between edge computing and cloud computing is not exclusive but complementary. This era represents a collaborative integration of both models, significantly enhancing the capabilities of edge computing in processing big data at the network's edge. This integration provides an optimal software and hardware support platform for the IoE. Nevertheless, the edge computing model still faces multifaceted challenges in managing data in the IoE era.

In February 2017, the Computing Community Consortium (CCC) released the "NSF Workshop Report on Grand Challenges in Edge Computing," [12] which discusses the main challenges of edge computing in areas such as applications, architecture, capabilities and services, and theoretical foundations of edge computing.

- **Application Challenges**: One of the main challenges in applying edge computing includes real-time processing and communication, security and privacy, incentives and profitability, adaptive application development, and the development and testing of application tools. Edge computing holds significant potential in applications such as video image analysis, virtual and augmented reality, deep learning, and intelligent connectivity and communication.
- **Architectural Challenges**: The architecture of edge computing encompasses several critical areas. Cage-level security ensures that massive data centers maintain high security unaffected by operator control through comprehensive hardware and software measures. Embracing approximation addresses the probabilistic nature of edge data processing due to inherent uncertainties in the data itself. The trade-off theory balances mobility, latency, capabilities, and

privacy to optimize system performance. Data provenance tracks the origin, usage, and intended users of large-scale data while preserving its integrity. Quality of Service (QoS) at the network edge guarantees end-to-end service quality of computing resources, fostering provider collaboration through mechanisms that define responsibility sharing, profit distribution, and resource utilization. Lastly, testbeds offer a cross-domain development environment equipped with appropriate standards and secure application programming interfaces (APIs) for edge computing applications.
- **Capabilities and Service Challenges**: This challenge includes resource naming, identification and discovery, standardized APIs, intelligent edge services, security and trust, and the edge service ecosystem. Efficiently utilizing edge computing resources largely depends on having a robust programming model or interface that makes it easier for developers to design and implement applications for the edge computing model. This support is crucial for the advancement of edge computing. A runtime system must provide support for the programming model at the higher level and effectively manage local resources at the lower level. It dynamically handles task partitioning and subtask deployment, ensuring smooth execution of each subtask at edge nodes and returning accurate results. In the edge computing model, although data storage and computation occur at the terminal, maintaining data security and privacy is essential. Effective privacy protection techniques must ensure that applications on edge nodes cannot access data from other applications and that external applications cannot access local data without proper authorization. The commercial model for edge computing will involve key players such as telecommunication operators, equipment providers, and edge device data producers, all integral to the edge computing business ecosystem. The commercial value of the edge computing industry encompasses edge service providers, data providers, and infrastructure builders. Data providers can fully leverage the value of local data, encouraging more edge terminals to join the edge computing framework.
- **Edge Computing Theory**: Edge Computing addresses the technical limitations of existing cloud computing technologies. However, refining the theoretical foundation and framework of edge computing is essential. This will provide better support for data processing in the IoE and promote the application of edge computing technologies across various critical fields.

The challenges mentioned earlier, along with others that have arisen throughout the development of edge computing, will be thoroughly explored in Chapter 5. Both industry and academia are actively working to overcome these challenges, and their efforts will be discussed in detail in Chapter 5.

2.4 Summary and Practice

2.4.1 Summary

This chapter begins with an introduction to distributed computing technologies, followed by an in-depth analysis of the fundamental concepts, models, and key technologies of edge computing. The edge computing model is based on a bidirectional computation flow, involving crucial technologies such as computation offloading, 5G/6G communication, new storage systems, lightweight libraries and kernels, and edge computing programming models. The relationship between edge computing and cloud computing is also examined, highlighting that edge computing is not intended to replace cloud computing but to complement and extend it; the two are mutually reinforcing. The integration of edge computing and cloud computing offers a more effective solution to the challenges of big data processing. Finally, the chapter discusses the inherent advantages of edge computing and provides a brief overview of the major challenges it faces in terms of application, architecture, capabilities and services, and edge computing theory.

2.4.2 Practice Questions

1) What are the key differences between distributed computing and centralized computing systems?
2) Describe the main functionalities of Hadoop's HDFS and how it supports distributed data storage.
3) Explain the basic concept of edge computing and its key characteristics.
4) Discuss the complementary relationship between edge computing and cloud computing.
5) Identify and explain two main challenges associated with implementing edge computing systems and propose potential solutions.

2.4.3 Course Projects

1) Conduct a comparative analysis of Hadoop MapReduce and Apache Spark for big data processing. By setting up Hadoop and Spark to perform a specific data-processing task, the performance can be measured and compared in terms of execution time, resource usage, scalability, and ease of development and debugging.
2) Design and implement a hybrid edge-cloud computing system for real-time data processing. The edge-based component can be set up in a basic edge

computing environment, and the cloud-based component can use a cloud service provider (e.g., AWS, Azure, and Google Cloud).

3) Explore the impact of latency on application performance. Set up a cloud-based application and measure its performance, then replicate the application in an edge computing environment and compare the results. Utilize tools like https://prometheus.io/ for monitoring and https://grafana.com/ for visualization to measure and compare latency.

Chapter 2 Suggested Papers

1 Rajkumar Buyya et al. "Cloud computing and emerging IT platforms: Vision, hype, and reality for delivering computing as the 5th utility". In: *Future Generation Computer Systems* 25. 6 (2009), pp. 599–616.

2 Jeffrey Dean and Sanjay Ghemawat. "MapReduce: Simplified data processing on large clusters". In: *Communications of the ACM* 51. 1 (2008), pp. 107–113.

3 Konstantin Shvachko et al. "The Hadoop distributed file system". In: *2010 IEEE 26th Symposium on Mass Storage Systems and Technologies (MSST)*. IEEE. 2010, pp. 1–10.

4 Lanyu Xu, Arun Iyengar, and Weisong Shi. "CHA: A caching framework for home-based voice assistant systems". In: *2020 IEEE/ACM Symposium on Edge Computing (SEC)*. IEEE. 2020, pp. 293–306.

5 Shanhe Yi et al. "LAVEA: Latency-aware video analytics on edge computing platform". In: *Proceedings of the 2nd ACM/IEEE Symposium on Edge Computing*. 2017, pp. 1–13.

6 Quan Zhang et al. "Firework: Big data sharing and processing in collaborative edge environment". In: *2016 4th IEEE Workshop on Hot Topics in Web Systems and Technologies (HotWeb)*. IEEE. 2016, pp. 20–25.

References

1 Yuan Ai, Mugen Peng, and Kecheng Zhang. "Edge computing technologies for Internet of Things: A primer". In: *Digital Communications and Networks* 4. 2 (2018), pp. 77–86.

2 Ganesh Ananthanarayanan et al. "Real-time video analytics: The killer app for edge computing". In: *Computer* 50. 10 (2017), pp. 58–67.

3 Apache. *Quarks*. https://quarks-edge.github.io/. Accessed: 2024-08-27.

4 Apache Software Foundation. *Apache Storm*. https://storm.apache.org/. Accessed: 2024-07-31.

5 Mark Baker, Rajkumar Buyya, and Domenico Laforenza. "Grids and Grid technologies for wide-area distributed computing". In: *Software: Practice and Experience* 32. 15 (2002), pp. 1437–1466.

6 Henri E Bal, Jennifer G Steiner, and Andrew S Tanenbaum. "Programming languages for distributed computing systems". In: *ACM Computing Surveys (CSUR)* 21. 3 (1989), pp. 261–322.

7 Paolo Bellavista, Antonio Corradi, and Cesare Stefanelli. "Mobile agent middleware for mobile computing". In: *Computer* 34. 3 (2001), pp. 73–81.

8 Philip A Bernstein. "Middleware: A model for distributed system services". In: *Communications of the ACM* 39. 2 (1996), pp. 86–98.

9 Aditya Bhardwaj and C Rama Krishna. "Virtualization in cloud computing: Moving from hypervisor to containerization—a survey". In: *Arabian Journal for Science and Engineering* 46. 9 (2021), pp. 8585–8601.

10 Rajkumar Buyya et al. "Cloud computing and emerging IT platforms: Vision, hype, and reality for delivering computing as the 5th utility". In: *Future Generation Computer Systems* 25. 6 (2009), pp. 599–616.

11 Fay Chang et al. "Bigtable: A distributed storage system for structured data". In: *ACM Transactions on Computer Systems (TOCS)* 26. 2 (2008), pp. 1–26.

12 Mung Chiang and Weisong Shi. *NSF Workshop Report on Grand Challenges in Edge Computing*. https://www.weisongshi.org/papers/shi16-nsfreport.pdf. Accessed: 2024-07-30.

13 Jeffrey Dean and Sanjay Ghemawat. "MapReduce: Simplified data processing on large clusters". In: *Communications of the ACM* 51. 1 (2008), pp. 107–113.

14 Marios D Dikaiakos et al. "Cloud computing: Distributed internet computing for IT and scientific research". In: *IEEE Internet Computing* 13. 5 (2009), pp. 10–13.

15 Domo. *Data Never Sleeps 11.0*. https://www.domo.com/learn/infographic/data-never-sleeps-11. Accessed: 2024-06-18.

16 Pedro Garcia Lopez et al. "Edge-centric computing: Vision and challenges". In: *ACM SIGCOMM Computer Communication Review* 45. 5 (2015), pp. 37–42.

17 Lars George. *HBase: The definitive guide*. O'Reilly Media, Inc., 2011.

18 Sanjay Ghemawat, Howard Gobioff, and Shun-Tak Leung. "The Google file system". In: *Proceedings of the 19th ACM Symposium on Operating Systems Principles*. 2003, pp. 29–43.

19 Chunye Gong et al. "The characteristics of cloud computing". In: *2010 39th International Conference on Parallel Processing Workshops*. IEEE. 2010, pp. 275–279.

20 Albert Greenberg et al. "The cost of a cloud: Research problems in data center networks". *ACM SIGCOMM Computer Communication Review* 39. 1 (2008), pp. 68–73.

21 Kyoungyoung Jee and Gang-Hoon Kim. “Potentiality of big data in the medical sector: Focus on how to reshape the healthcare system”. In: *Healthcare Informatics Research* 19. 2 (2013), pp. 79–85.

22 Danny B Lange. “Mobile objects and mobile agents: The future of distributed computing?” In: *European Conference on Object-Oriented Programming*. Springer. 1998, pp. 1–12.

23 Loren Lugosch et al. “Speech model pre-training for end-to-end spoken language understanding”. In: *arXiv preprint arXiv:1904.03670* (2019).

24 James Manyika et al. *Big data: The next frontier for innovation, competition, and productivity*. McKinsey Global Institute, 2011.

25 Rino Micheloni. “Solid-state drive (SSD): A nonvolatile storage system”. In: *Proceedings of the IEEE* 105. 4 (2017), pp. 583–588.

26 Flávio Ramalho and Augusto Neto. “Virtualization at the network edge: A performance comparison”. In: *2016 IEEE 17th International Symposium on A World of Wireless, Mobile and Multimedia Networks (WoWMoM)*. IEEE. 2016, pp. 1–6.

27 Konstantin Shvachko et al. “The Hadoop distributed file system”. In: *2010 IEEE 26th Symposium on Mass Storage Systems and Technologies (MSST)*. IEEE. 2010, pp. 1–10.

28 Ashish Singh and Kakali Chatterjee. “Securing smart healthcare system with edge computing”. In: *Computers & Security* 108 (2021), p. 102353.

29 Karolj Skala et al. “Scalable distributed computing hierarchy: Cloud, fog and dew computing”. In: *Open Journal of Cloud Computing (OJCC)* 2. 1 (2015), pp. 16–24.

30 Blesson Varghese et al. “Challenges and opportunities in edge computing”. In: *2016 IEEE International Conference on Smart Cloud (SmartCloud)*. IEEE. 2016, pp. 20–26.

31 VFS Global. *Building Privacy into 5G Technology cannot be an afterthought*. https://www.vfsglobal.com/en/individuals/insights/building-privacy-into-5G-technology-cannot-be-an-afterthought.html. Accessed: 2024-08-27.

32 Tom White. *Hadoop: The definitive guide*. O’Reilly Media, Inc., 2012.

33 Qishi Wu et al. “On computing mobile agent routes for data fusion in distributed sensor networks”. In: *IEEE Transactions on Knowledge and Data Engineering* 16. 6 (2004), pp. 740–753.

34 Lanyu Xu, Arun Iyengar, and Weisong Shi. “CHA: A caching framework for home-based voice assistant systems”. In: *2020 IEEE/ACM Symposium on Edge Computing (SEC)*. IEEE. 2020, pp. 293–306.

35 Shanhe Yi et al. “LAVEA: Latency-aware video analytics on edge computing platform”. In: *Proceedings of the 2nd ACM/IEEE Symposium on Edge Computing*. 2017, pp. 1–13.

36 Matei Zaharia et al. "Spark: Cluster computing with working sets". In: *2nd USENIX Workshop on Hot Topics in Cloud Computing (HotCloud 10)*. 2010.

37 Quan Zhang et al. "Firework: Big data sharing and processing in collaborative edge environment". In: *2016 4th IEEE Workshop on Hot Topics in Web Systems and Technologies (HotWeb)*. IEEE. 2016, pp. 20–25.

3

Architecture and Components of Edge Computing*

In Chapters 1 and 2, we explored why edge computing has become a necessity in the modern digital landscape and examined its foundational principles. Now, as we delve into the architecture and components of edge computing, we will uncover how this innovative approach is structurally and functionally different from traditional centralized models. This chapter will provide a comprehensive look at the critical infrastructure, computing models, and networking considerations that form the backbone of edge computing.

3.1 Edge Infrastructure

To fully understand the capabilities and potential of edge computing, it is essential to first examine its underlying infrastructure. Unlike centralized models that rely on a few powerful data centers, edge computing is built on a distributed network of devices closer to the data source. In this section, we will explore how edge infrastructure is uniquely constructed to meet the demands of low-latency, high-bandwidth applications, and the various grades/layers that define its architecture.

3.1.1 Introduction to Edge Computing Architecture

Edge computing is a network of decentralized computing systems that process and analyze data near its origin, rather than relying on centralized data centers. The key components of edge infrastructure include edge devices, which can be physical devices such as smartphones, Internet of Things (IoT) devices, or edge

*This chapter is contributed by Shihong Hu and Xingzhou Zhang.

Edge Computing: Systems and Applications, First Edition. Lanyu Xu and Weisong Shi.

Companion Website: www.wiley.com/go/xu/computing

servers; edge data centers, which are smaller, localized data centers that provide additional processing power; and communication networks, which facilitate the transfer of data between edge nodes and the central cloud.

Edge nodes are equipped with processing capabilities and are responsible for executing tasks locally, reducing the need to send data to the cloud. Edge data centers serve as local hubs that can handle more intensive processing tasks and provide a buffer for data before it is sent to the central cloud or other edge nodes. The communication networks are critical for ensuring that data can be transferred efficiently and securely between different components of the edge infrastructure.

3.1.1.1 Differentiation from Centralized Models

The primary distinction between edge infrastructure and centralized models lies in the distribution of computational resources and data processing. Centralized models typically involve large data centers that are geographically distant from the end-users and devices generating data. These models can become bottlenecks due to the sheer volume of data that needs to be transmitted over long distances, leading to increased latency and potential network congestion. In contrast, edge infrastructure reduces these issues by processing data locally or regionally, minimizing the amount of data that needs to be sent to the cloud. This approach not only decreases latency but also enhances the overall efficiency and responsiveness of applications and services.

The construction goals of edge computing are as follows:

- **Enhancing real-time capability:** The scenario of pervasive connectivity demands applications with an exceptionally high requirement for real-time performance. In the conventional cloud computing paradigm, applications transmit data to remote cloud centers and subsequently request processed results, thereby augmenting the system's latency. Edge computing is poised to augment the immediacy of data processing. For instance, in the case of autonomous vehicle applications, vehicles in rapid motion necessitate reaction times on the order of milliseconds. Any exacerbation of system latency due to network issues could precipitate grave outcomes.
- **Diminishing bandwidth demand:** Edge devices incessantly generate substantial volumes of data, and the transmission of this data in its entirety to the cloud imposes a significant strain on network bandwidth. A case in point is the Boeing 787, which produces over 5 gigabytes of data per second [13]. However, the bandwidth available between the aircraft and satellite links is inadequate to facilitate real-time data transmission.
- **Mitigating energy consumption:** Data centers are prodigious consumers of energy. Research conducted by Sverdlik [40] indicates that by the year 2020, the cumulative energy consumption of all data centers in the United States will have

escalated by 4%, amounting to 73 billion kilowatt-hours. As the proliferation of user applications and the magnitude of data processed continue to escalate, energy consumption is poised to emerge as a constraint impeding the growth of cloud computing centers. The distributed processing characteristic of edge computing can attenuate the energy footprint of data centers.

- **Safeguarding data security and privacy:** Data within the interconnected milieu is intimately connected to users' lives. For example, the installation of indoor smart network cameras in many homes means that video data transmitted to the cloud heightens the risk of privacy breaches. With the advent of the European Union's General Data Protection Regulation (GDPR) [14], concerns regarding data security and privacy have acquired increased significance for cloud computing enterprises. Edge computing offers robust mechanisms for the protection of data security and privacy.

3.1.2 Different Grades/Layers of Edge

Edge computing is a continuum, which represents a spectrum of computational paradigms extending from centralized cloud infrastructures to the very periphery where data is generated. As the architecture of edge computing matures, it has evolved to encompass a multitiered structure of edges, each with distinct characteristics and capabilities. As shown in Figure 3.1, we will analyze three primary categories of edge computing: on-premises edge, network edge, and data center edge, elucidating their unique attributes and potential applications.

3.1.2.1 On-premises Edge

On-premises edge computing refers to the processing of data at the physical location of the user or data source. This type of edge computing is typically applied in scenarios that require real-time data processing and rapid response. For example, in smart venues, on-premises edge data centers can analyze audience traffic, security monitoring, and environmental conditions in real time, thereby providing a safer and more personalized experience. In the field of intelligent connected vehicles, in-vehicle data-processing units can process data from sensors to achieve functions such as autonomous driving, collision prevention, and vehicle status monitoring. In public safety video processing, cameras typically have some data preprocessing capabilities. These cameras are characterized by low power consumption and limited computational resources, but they generally can filter data to reduce bandwidth usage.

The advantages of on-premises edge computing are manifold. The reduced latency allows for faster, more efficient data processing, which is essential for time-sensitive applications. The high reliability of local processing ensures that operations can continue even in the event of network disruptions.

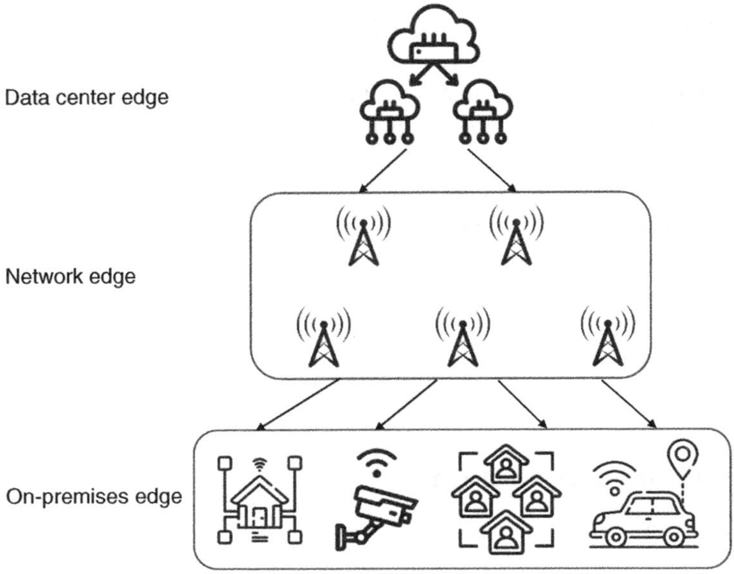

Figure 3.1 Three layers of edge.

Moreover, enhanced data privacy is a significant benefit, as sensitive information is processed and stored locally, minimizing the risk of data breaches.

3.1.2.2 Network Edge

The network edge expands the reach of edge computing beyond the immediate location of data generation by leveraging the infrastructure of telecommunications operators. This category of edge computing is integral to the functioning of mobile networks, where base stations equipped with edge computing capabilities can handle data-processing tasks that were traditionally performed in central data centers.

The network edge is particularly transformative for mobile network optimization, enabling the delivery of high-quality services such as video streaming and online gaming with reduced latency. The deployment of edge computing at the base stations allows for localized content caching, which accelerates content delivery and enhances user experience. In the area of IoT, the network edge is instrumental in managing the vast array of connected devices. By processing data at the edge of the network, the load on central servers is reduced, and the responsiveness of IoT applications is improved. This is particularly beneficial for smart city initiatives, where numerous devices, from traffic lights to environmental sensors, require real-time data processing and analytics.

3.1.2.3 Data Center Edge

The data center edge represents a strategic extension of traditional data center capabilities to the edge of the network. This category of edge computing is designed to bridge the gap between centralized data processing and the localized needs of on-premises and network edges. Data center edge computing is particularly adept at handling large-scale data-processing tasks that require high computational power and storage capacity. By deploying resources closer to the users, data center edge computing ensures that high-demand applications, such as online gaming platforms and social media networks, can operate with minimal latency and maximum performance.

One of the key applications of data center edge computing is in the multiregion deployment strategies of cloud service providers. By establishing edge computing nodes in various geographical locations, these providers can offer consistent service quality and response times to users across the globe, regardless of their location. Moreover, the data center edge is a critical component in the realm of big data and machine learning. The proximity to end-users enables real-time data analysis and inference, providing enterprises with valuable insights that can inform strategic decision-making and operational adjustments.

3.1.3 Capabilities of Edge Infrastructure

Edge infrastructure represents a pivotal technological advancement that facilitates the decentralization of computing resources. By bringing computation and data storage closer to the source of data generation, edge infrastructures enhance the efficiency, speed, and security of data-processing operations. This section aims to provide more analysis of the capabilities of edge infrastructure: data process, cache and storage, communication, and content delivery networks (CDNs).

3.1.3.1 Data Process

At the heart of edge infrastructure lies the capability for data processing, which encompasses a range of operations critical to the preprocessing of data. The data process includes collection, filtering, cleansing, transformation, and aggregation. Collection is the initial step, where data from various sources is gathered. Filtering then follows, allowing the system to select relevant data points according to predefined criteria. Cleansing ensures the removal of corrupt or irrelevant data, thereby maintaining data integrity. Transformation adjusts the format or structure of the data to fit the requirements of downstream applications. Finally, aggregation consolidates data into a summarized form, making it more manageable and insightful for analysis. Data process in edge infrastructure enables real-time analytics and decision-making, which is crucial for applications such as predictive maintenance in industrial IoT, real-time traffic management in smart cities, and instantaneous decision-making in financial trading systems.

3.1.3.2 Cache and Storage

Cache and storage is the foundational capability that supports the data lifecycle within edge infrastructures [55]. It involves the temporary or permanent storage of data, which can be historical or recently generated. The temporary storage, or caching, allows for quick access and retrieval of frequently used data, thereby reducing latency and improving response times. On the other hand, the storage of historical data is vital for applications that require trend analysis, long-term planning, or compliance with data retention policies. The design of cache and storage systems in edge infrastructures must consider factors such as data durability, accessibility, and security. The use of distributed file systems, object storage, and databases that are optimized for edge environments ensures that data can be stored, managed, and retrieved efficiently.

3.1.3.3 Communication

Communication is the glue that binds the components of edge infrastructures and cloud servers together. It facilitates the forwarding of data to other nodes within the edge network or to central servers for further processing or analysis [31]. The efficiency of communication protocols directly impacts the performance of edge applications, particularly in scenarios that demand high throughput and low latency, such as real-time video streaming or remote healthcare services. Edge infrastructures will support robust communication mechanisms that can handle diverse data types and volumes. This includes the use of both wired and wireless communication technologies, as well as the implementation of communication protocols that are resilient to network failures and capable of adapting to varying network conditions.

3.1.3.4 CDNs

CDN represents a specialized application of edge infrastructure capabilities, which is designed to optimize the delivery of content to end-users by replicating it across multiple edge nodes. This distribution reduces the latency associated with content retrieval and ensures a high availability of resources, even during peak traffic periods. The role of CDN in edge infrastructures extends beyond mere content caching. It involves sophisticated algorithms for content routing, load balancing, and dynamic content optimization. CDNs also play a critical role in enhancing the security of content delivery through techniques such as distributed denial-of-service (DDoS) protection and secure content delivery.

3.1.4 New Progress of Edge Computing Architecture

In recent years, the architecture of edge computing has seen further development, and this chapter outlines several typical directions of evolution.

3.1.4.1 Edge Collaborative Consortium Architecture

The edge computing paradigm aspires to extend the network service advantages of cloud data centers toward the network edge, thereby bringing services closer to users and computations closer to the source of data, offering faster service response times. To achieve this, edge infrastructure providers (EIPs) need to deploy computing and storage resources at appropriate locations within the access network and allow edge service providers (ESPs) to leverage the edge layer resources provided by EIPs to deliver critical services to users. Compared to the cloud computing model, edge nodes face issues of limited computing, storage, and bandwidth resources, and the large-scale construction of edge nodes also entails high construction and maintenance costs. Therefore, EIPs are more inclined to establish a series of small-scale, private edge computing environments to meet the specific needs of users.

3.1.4.2 Computing-Networking Integration Architecture

The computing-networking integration architecture has evolved from networking, cloud networking to computing-networking, where networking is the foundation, creating lossless and deterministic network connections; cloud networking is a further advancement in networking and cloudification; and computing-networking achieves trustworthy, efficient, on-demand, low-cost, and flexible computing services. The development of computing-networking integration can be summarized into three important stages: intra-data center computing-networking integration, cloud-networking integration, and cloud-edge-end computing-networking integration. Currently, the academic community in China has proposed the concept of computation, which is a form of computing-networking integration. Sky computing was proposed by UC Berkeley (UCB) [39], which unites multiple data centers at the cloud edge and end for collaborative optimization, also representing a form of computing power integration.

3.1.4.3 Edge-Native Architecture

The introduction of the edge native concept has brought more agile development architectures, simplified operational configurations, and new value propositions to edge layer applications. Edge native refers to the architectural design of applications with the deployment on the edge network as the target to fully leverage edge capabilities. It is two sides of the same coin with cloud native, reflecting the continuous shift of the information and communication technology (ICT) industry's focus toward the edge. Compared to cloud native, edge native also considers characteristics such as rapid deployment, continuous delivery, and the elasticity of shielding underlying implementations, but in response to the unique complex networking forms, limited resources, and the diversity of computing and communication hardware at the edge, edge native places greater emphasis on features such as integrated computation and communication, lightweight, support for heterogeneous devices, and autonomous offline edge capabilities.

3.1.5 Open Questions

3.1.5.1 Computing Ability

In recent years, large language models and large video models have made good progress in many fields, and researchers have begun to study how to implement them in edge scenarios. However, the self-attention mechanism [43] at the core of large language/video models leads to computation and memory usage growing quadratically with the number of patches. This makes it difficult to deploy large language/video models on the edge with limited computational power and memory capacity. For example, a typical Swin-L [28] model, contains approximately 197 million parameters and requires about 104 GFLOPs for a single forward propagation when processing a 384x384 resolution image. In contrast, the NVIDIA Jetson Nano, a common edge computing device, has a theoretical computational power of only 472 GFLOPs, allowing it to process up to about four image frames per second under ideal conditions. What is more, due to practical factors such as system overhead, memory bandwidth limitations, and computational efficiency, the actual processing speed is often much lower than the theoretical value. Therefore, we should focus on how to make full use of the computing resources of edge devices at different layers to meet the computing power requirements of large AI models.

3.1.5.2 Programmability

Programming models facilitate the rapid onboarding of developers in the creation of application products, thus hastening the evolution of their respective domains. In the area of cloud computing, user programs are authored and compiled on the target platform, subsequently being executed on cloud servers with the underlying infrastructure remaining opaque to the user. Amazon's Lambda service, for example, leverages such a programming model, enabling users to operate code without the prerequisite of preconfiguring or managing servers, significantly enhancing user convenience. Nonetheless, the paradigm of edge computing diverges markedly from that of cloud computing, exhibiting attributes of elastic management, collaborative execution, and environmental heterogeneity. Edge computing necessitates the segmentation of applications, the dispersion of data, and the distribution of resources. As a result, conventional programming models fall short in meeting the demands of edge computing.

In the domain of edge computing, the majority of devices constitute heterogeneous computing platforms, each with its unique runtime environment and data sets. Furthermore, the resources available on edge devices are relatively limited, presenting considerable difficulties in the deployment of user applications within edge computing environments. Hence, there is an imperative to explore an innovative programming model, analogous to the MapReduce paradigm in the big data sphere, which can offer a unified and succinct programming methodology tailored for a multitude of applications, thereby driving the technological progression of edge computing to new heights.

3.2 Edge Computing Models

With the foundation of edge infrastructure laid out, we now turn our attention to the various computing models that drive edge computing. These models range from simple, device-level processing to complex, collaborative systems that involve multiple nodes working together.

3.2.1 Overview and Definitions

Edge computing is revolutionizing how data is processed and managed by bringing computation and storage closer to the data source. Two prominent paradigms in this domain are mobile edge computing (MEC) and cloudlet computing (Figure 3.2). Specially, Mobile Edge Computing and multi-access edge computing [7] are essentially the same concept, with the term "Multi-access Edge Computing" being an evolution of "Mobile Edge Computing." MEC, standardized by the European Telecommunications Standards Institute (ETSI) [19], integrates cloud services at the edge of the network, enabling low-latency and high-bandwidth applications. It provides a platform for deploying applications and services that require real-time processing, such as augmented reality (AR), autonomous driving, and industrial automation. On the other hand, cloudlet computing [30], a concept introduced by Carnegie Mellon University, focuses on providing localized, powerful computing resources near mobile devices. Cloudlets serve as small-scale data centers that offer cloud-like capabilities with minimal latency. They are particularly suited for offloading computation-intensive tasks from mobile devices, thereby enhancing performance and extending battery life.

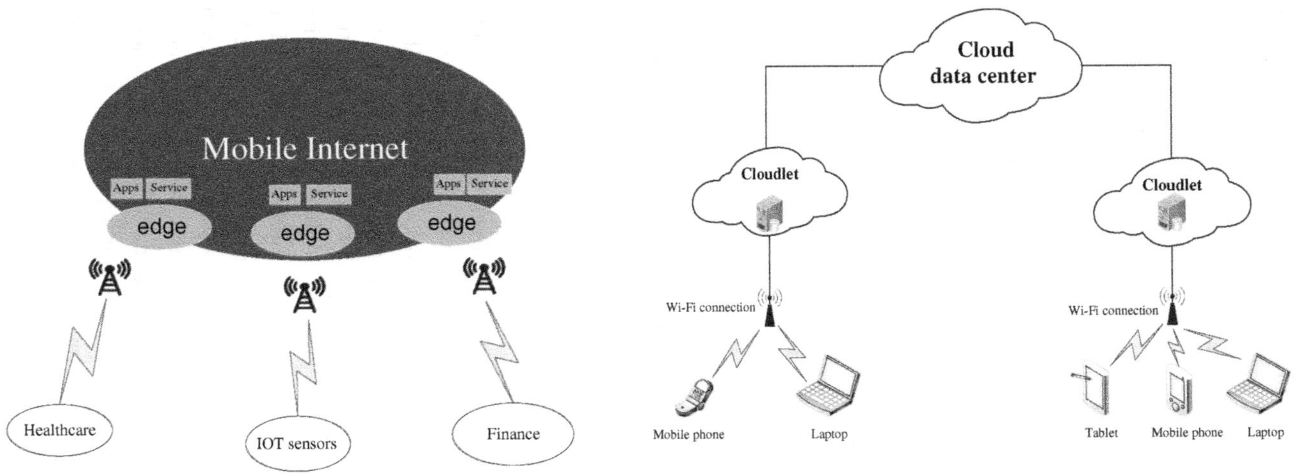

Figure 3.2 "MEC" and "cloudlet computing."

While MEC and cloudlet computing share the common goal of bringing computation closer to the edge, they differ significantly in architecture, deployment, and use cases. MEC is tightly integrated with the mobile network infrastructure, leveraging the existing cellular network to provide seamless connectivity and service continuity. This integration makes MEC ideal for telecom operators aiming to offer value-added services and optimize network performance [29]. In contrast, cloudlet computing emphasizes flexibility and decentralization [2]. Cloudlets can be deployed independently of the network infrastructure, providing localized computing resources that can be easily scaled and managed. This independence allows for rapid deployment in diverse environments, from urban areas to remote locations, making cloudlets highly adaptable to varying application requirements. Furthermore, MEC's close association with telecom infrastructure ensures robust security and privacy controls, which are crucial for applications in healthcare, finance, and other sensitive sectors. Cloudlet Computing, while also capable of maintaining high-security standards, often relies on end-to-end encryption and local data processing to protect user data.

As the edge computing landscape evolves, advanced models are emerging to address the limitations of current paradigms and harness the full potential of edge technologies. The rise of 5G networks is poised to further amplify the capabilities of edge computing models. The ultralow latency and high bandwidth of 5G will enable more sophisticated applications and seamless integration of MEC and cloudlet computing. These advancements will pave the way for innovative services and applications that were previously unattainable due to latency and bandwidth constraints. An emerging trend within this context is the development of collaborative edge computing models. These models emphasize the cooperation and coordination among multiple edge nodes to optimize resource utilization, enhance scalability, and improve fault tolerance. By leveraging collaborative frameworks, edge nodes can dynamically share workloads, balance traffic, and provide redundancy, thus ensuring more reliable and efficient service delivery. In conclusion, the future of edge computing lies in the synergistic integration of various models and technologies. MEC and cloudlet computing, with their unique strengths, are foundational to this ecosystem. As technology progresses, the development of advanced models, including collaborative edge computing, and the incorporation of AI and 5G will continue to drive the evolution of edge computing, unlocking new possibilities and transforming industries across the globe.

3.2.2 Collaborative Edge Computing Models

3.2.2.1 Edge-to-Edge Collaboration

Edge-to-edge collaboration involves direct interaction and coordination between edge nodes to enhance performance, reliability, and scalability [45]. This model

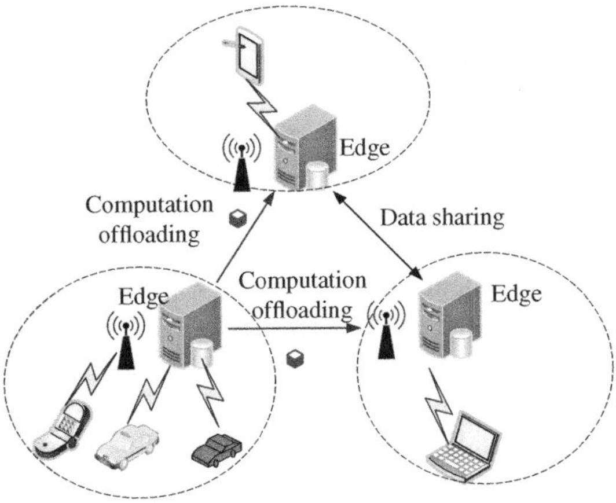

Figure 3.3 Edge-to-edge collaboration.

is essential for distributed applications requiring real-time data processing and resource sharing across multiple edge locations. In edge-to-edge collaboration, edge nodes communicate directly with each other to share data, balance workloads, and provide redundancy, as shown in Figure 3.3. This collaboration is facilitated through decentralized protocols and frameworks that enable efficient resource allocation and management without relying on a central authority. By leveraging local interconnections, edge nodes can reduce latency and improve data processing speeds. Key applications of edge-to-edge collaboration include smart city infrastructure, autonomous vehicle networks, and industrial IoT systems. In these scenarios, edge nodes work together to process vast amounts of data generated by sensors and devices, enabling real-time decision-making and actions. For example, Li et al. [27] proposes the data sharing scheme among intelligent connected vehicles to ensure safe driving, which is the typical scenario of edge-to-edge collaboration.

3.2.2.2 Edge-to-Device Collaboration

Edge-to-device collaboration focuses on the interaction between edge nodes and end-user devices, such as smartphones, wearables, and IoT sensors, as depicted in Figure 3.4. This collaboration model is pivotal in offloading computation-intensive tasks from devices to nearby edge nodes, thereby improving device performance and battery life. Prominent use cases include AR applications, where real-time processing is essential for rendering graphics and ensuring smooth user experiences [12]. Similarly, in healthcare, wearable devices can offload data-processing tasks to edge nodes, enabling continuous health monitoring and immediate response to critical health events [38]. Another example is in smart homes, where edge nodes

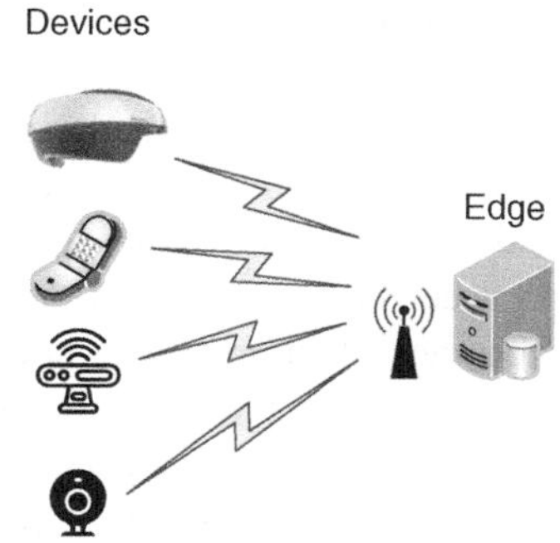

Figure 3.4 Edge-to-device collaboration.

manage and process data from various connected devices to optimize energy usage and enhance security [51].

3.2.2.3 Edge-to-Cloud Collaboration

Edge-to-cloud collaboration involves the integration of edge computing resources with centralized cloud infrastructures. This hybrid model leverages the strengths of both edge and cloud computing, providing a balanced approach to data processing and storage, as shown in Figure 3.5. The primary benefits of edge-to-cloud collaboration include enhanced scalability, as cloud resources can be used to handle peak loads and extensive data storage. Additionally, this model supports advanced analytics and machine learning applications, where large datasets can be processed in the cloud while critical, time-sensitive computations are performed at the edge [50]. For example, Wu et al. [48] demonstrated that the hybrid human-AI scheme, enabled by edge-cloud collaboration, offers a

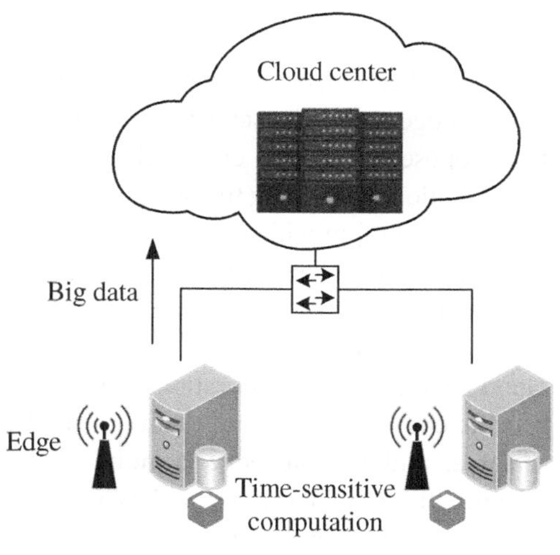

Figure 3.5 Edge-to-cloud collaboration.

promising solution for enhancing video services. By combining the strengths of edge and cloud computing with human insights and AI capabilities, the proposed approach can significantly improve video service quality and user experience. However, integrating edge and cloud resources presents challenges, including data consistency, latency, and security. Ensuring seamless data synchronization between edge nodes and cloud servers is crucial to maintaining the integrity and accuracy of information. Latency issues must be addressed to provide timely responses for real-time applications. Furthermore, robust security measures are required to protect data as it moves between edge nodes and the cloud. For example, the study in [20] presented an innovative hybrid DDPG-D3QN approach for intelligent resource allocation in edge-cloud collaborative networks. By combining the strengths of DDPG and D3QN, the proposed method achieves superior performance in optimizing resource utilization, reducing latency, and minimizing costs, thereby enhancing the overall efficiency of edge-cloud systems.

3.2.2.4 Cloud-Edge-Device Collaboration

Cloud-edge-device collaboration represents a holistic approach where tasks are distributed across cloud servers, edge nodes, and end devices based on their computational requirements and latency sensitivity. This layered model ensures optimal resource utilization and enhances the overall system performance. In this collaborative framework, cloud servers handle large-scale data processing and long-term storage, edge nodes perform intermediate processing and provide low-latency responses, and devices focus on data collection and immediate user interactions, as depicted in Figure 3.6. By strategically distributing tasks, this model leverages the unique strengths of each layer, providing a robust solution for complex and dynamic applications. Peng et al. [34] presented an end-edge-cloud collaborative computation offloading framework designed for multiple mobile users in a heterogeneous edge-server environment. By leveraging the strengths of both edge and cloud servers and addressing the heterogeneity of edge servers, the proposed approach effectively reduces task completion time and energy consumption, enhancing the overall performance of mobile applications. Besides, Wang et al. [46] provided a comprehensive overview of the current landscape of end-edge-cloud collaborative computing for deep learning.

Collaborative edge computing models play a critical role in the evolving landscape of edge computing. Edge-to-edge, edge-to-device, edge-to-cloud, and cloud-edge-device collaborations each offer unique advantages and address specific challenges, making them essential components in developing efficient, scalable, and reliable edge computing systems. As technology continues to advance, these collaborative models will enable innovative applications and drive the next wave of digital transformation.

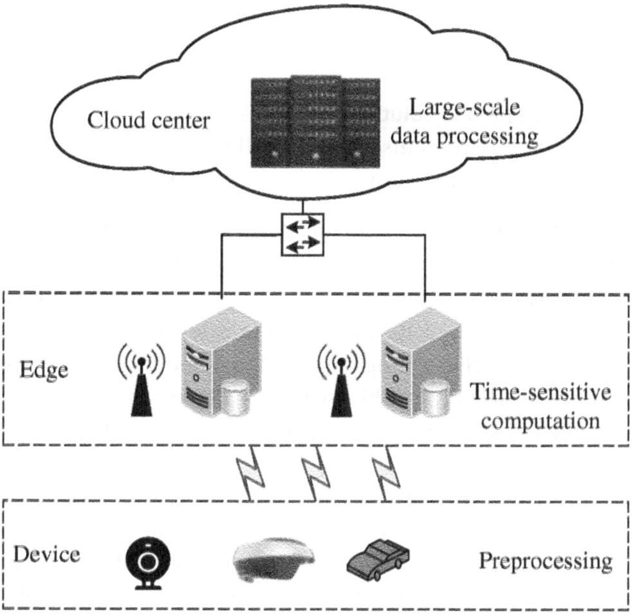

Figure 3.6 Cloud-edge-device collaboration.

3.2.3 Choosing the Right Model

3.2.3.1 Factors to Consider

When selecting an edge computing model, several factors must be taken into account to ensure that the chosen solution aligns with business needs and technical constraints.

- **Business needs:** Understanding the specific requirements and goals of the business is crucial. Edge computing models should be evaluated based on their ability to support these objectives. For instance, industries such as healthcare and finance, which demand high levels of data privacy and security, may benefit more from models that provide robust encryption and localized data processing. In contrast, applications in smart cities or retail, where real-time data analytics and responsiveness are critical, might prioritize models with low latency and high availability.
- **Technical constraints:** Technical limitations and infrastructure capabilities also play a significant role in determining the appropriate edge computing model. Factors such as existing network architecture, available bandwidth, latency requirements, and computational power at the edge nodes should be considered. For example, environments with limited network connectivity

might favor cloudlet computing due to its ability to operate independently of the central cloud.

3.2.3.2 Implementation Challenges and Solutions

Deploying edge computing systems presents unique challenges that need to be addressed to ensure successful implementation.

- **Addressing deployment challenges:** One of the primary challenges is managing the distributed nature of edge nodes. Effective orchestration and management tools are necessary to monitor, update, and maintain these nodes. Solutions like Kubernetes, and its edge-specific extension KubeEdge, provide robust frameworks for orchestrating containerized applications across edge and cloud environments. They enable seamless deployment, scaling, and management of applications, thereby simplifying the complexity of edge computing deployments.
- **Security and privacy:** Ensuring data security and privacy at the edge is another critical challenge. Implementing comprehensive security measures, including encryption, access control, and regular security updates, is essential. Solutions should be designed to protect data both in transit and at rest, ensuring compliance with regulatory standards and safeguarding against potential breaches.

3.2.3.3 Typical Edge Computing Systems and Models

Several edge computing systems have been developed to address these challenges and provide effective models for various use cases.

- **KubeEdge:** KubeEdge is an open-source platform that extends Kubernetes capabilities to edge nodes [49]. It enables the deployment and management of containerized applications at the edge, providing a scalable and flexible solution for edge computing. KubeEdge supports a wide range of use cases, from industrial IoT to smart cities, by facilitating real-time data processing and efficient resource utilization [44].
- **K3s:** K3s is a lightweight Kubernetes distribution designed specifically for resource-constrained environments such as edge computing [35]. It aims to deliver a quick, straightforward, and efficient way to establish a highly available and fault-tolerant cluster across a group of nodes focused on low–end application areas. K3s demonstrates minimal disk usage, likely due to its use of SQLite database. It also exhibits performance benefits in most operations when compared to other Kubernetes, and is suitable for starting new nodes and integrating them into the cluster [4].
- **MicroK8s:** MicroK8s is another lightweight, single-node Kubernetes distribution designed for edge computing. It is easy to install and manage, therefore suitable for edge scenarios [5].

- **Azure IoT Edge:** Azure IoT Edge is another prominent solution that allows for the deployment of cloud workloads to edge devices [23]. It supports a variety of programming languages and can run AI and analytics workloads locally, reducing latency and bandwidth usage. Azure IoT Edge is particularly useful in scenarios where intermittent connectivity is a concern, as it ensures continued operation even when the connection to the cloud is lost.
- **AWS IoT Greengrass:** AWS IoT Greengrass brings cloud capabilities to local devices, enabling them to collect and analyze data closer to the source [25]. It supports machine learning inference, device messaging, and data sync with AWS cloud, making it suitable for complex, data-intensive applications. Greengrass's ability to operate offline and sync when connectivity is restored makes it a robust solution for remote and mobile environments.

3.2.3.4 New Progress of Edge Model

The field of edge computing is continuously evolving, with significant advancements in edge federation and edge AI models.

- **Edge federation:** Edge federation is an emerging concept where multiple edge nodes, often managed by different entities, collaborate to provide a unified computing platform. This approach enhances resource utilization, ensures better load balancing, and provides redundancy, improving overall system resilience and performance. Edge federation is particularly beneficial in scenarios requiring high availability and robust fault tolerance, such as smart grids, autonomous transportation systems, and large-scale IoT deployments. For example, Cao et al. [6] presents edge federation, an integrated service provisioning model for the edge computing paradigm. It aims to establish a cost-efficient platform for edge infrastructure providers (EIPs) and offer end users and edge service providers a transparent resource management scheme by seamlessly integrating individual EIPs as well as clouds.
- **Edge AI models:** Integrating AI at the edge, or edge AI, is a rapidly advancing field that enables real-time data analysis and decision-making at the source of data generation [37]. Edge AI models leverage the computational capabilities of edge devices to run machine learning algorithms locally. This reduces the need for constant data transmission to the cloud, thereby lowering latency and bandwidth usage. Edge AI is instrumental in applications such as predictive maintenance, real-time anomaly detection, and enhanced user experiences in AR/VR environments. The combination of edge computing and AI opens up new possibilities for smart, autonomous systems capable of making informed decisions without relying on central cloud resources. Federated learning is a key component of edge AI that enables decentralized machine learning [26]. In traditional AI models, data is collected and sent to a central server for processing and model training. Federated learning, however, allows edge devices to

collaboratively train models without sharing raw data. Each device processes its local data and sends only model updates to a central server, which aggregates the updates to improve the global model. This approach enhances privacy, reduces bandwidth usage, and leverages the computational power of edge devices [53].

3.2.4 Open Questions

As edge computing continues to evolve, several open questions remain, particularly concerning the continuity and unity of edge computing models.

3.2.4.1 Continuity

The edge computing model is envisioned as a continuum, seamlessly integrating devices, edge nodes, and cloud resources. However, current models are often dominated by single layers, with specific tasks relegated either to the cloud, edge, or device level. Achieving continuity in computing is a significant challenge that requires:

- **Dynamic orchestration:** Developing advanced orchestration frameworks that can dynamically allocate tasks across the continuum based on real-time needs and resource availability. These frameworks must be capable of fluidly shifting workloads between the cloud, edge, and devices without disrupting services.
- **Interoperability standards:** Establishing common standards and protocols that ensure interoperability between different layers of the continuum. This includes standardized APIs and communication protocols that enable seamless interaction and data exchange across diverse systems and platforms.
- **Adaptive algorithms:** Implementing adaptive algorithms that can intelligently distribute computational tasks based on factors such as latency requirements, computational load, and network conditions. These algorithms must be able to learn and evolve, optimizing task distribution to maintain continuity in service delivery.
- **Edge-native applications:** Encouraging the development of edge-native applications designed to leverage the full spectrum of the edge continuum. These applications should be capable of dynamically adjusting their computational strategies to utilize resources efficiently across different layers.

3.2.4.2 Unity

Edge computing currently relies heavily on specific scenarios, tailored to particular applications or industries. This scenario-based approach, while effective for targeted use cases, raises questions about the potential for unity across different scenarios. Achieving unity in edge computing involves:

- **Unified frameworks:** Creating unified frameworks that support a wide range of applications and use cases. These frameworks should provide the necessary tools and abstractions to accommodate diverse requirements while maintaining a consistent underlying architecture.
- **Cross-domain collaboration:** Promoting collaboration across different industries and sectors to develop shared solutions and best practices. This collaborative approach can lead to the creation of versatile edge computing platforms that are applicable to multiple scenarios.
- **Modular design:** Adopting a modular design philosophy for edge computing systems, where components can be easily adapted or replaced to suit different scenarios. This modularity ensures that core functionalities remain consistent while allowing for customization to meet specific needs.
- **Interdisciplinary research:** Encouraging interdisciplinary research that combines insights from various fields, such as computer science, telecommunications, and industrial engineering. This research can drive the development of innovative solutions that bridge the gap between different edge computing scenarios.

3.3 Networking in Edge Computing

Having discussed the infrastructure and computing models, it's time to consider one of the most critical aspects of edge computing—networking. The integration of networking within edge computing is crucial to achieving seamless data processing and communication across distributed environments.

3.3.1 Introduction and Development Process of Edge Computing-Network Integration

In edge computing multilayer architecture, how the network communication between different layers is realized is shown in Figure 3.7. The data centers in the cloud are connected by the inter-data center network. This network supports communication between multiple data centers, enabling distributed processing and redundancy. To provide cloud-like service to end users, edge devices must be connected to the cloud center and end users through network infrastructures, referred to as edge networking. Edge computing devices communicate with each other and with other network components through edge networking. This layer facilitates the direct exchange of data between edge devices, ensuring low-latency communication for time-sensitive applications. The edge nodes are connected to the cloud center through the core/metro networks. Core-metro networks and the internet support communication between cloud data centers, edge devices,

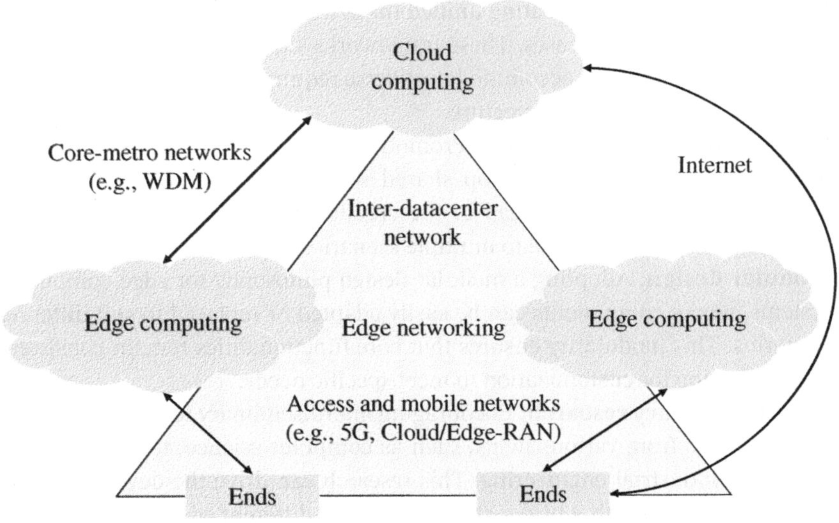

Figure 3.7 Edge computing and networking.

and end users, providing the necessary bandwidth and speed for large-scale data transfer [54]. Edge networks connect to access and mobile networks, such as 5G and Cloud/Edge-RAN, to extend the reach of computing resources to end users. At the same time, end devices can send data to and receive data from the edge computing layer, enabling real-time applications and services.

Recently, the convergence of networking and edge computing has been accelerated by the trend of network cloudification, highlighting the pivotal role of networking in cutting-edge computing technologies [11]. This convergence promotes a comprehensive vision that integrates resource and function management across both network and edge systems, enabling unified provisioning of services that span network and cloud/edge environments [10]. Traditionally, networking functions as the backbone for data transmission, resource sharing, and connectivity in edge computing environments. It ensures efficient data flow between devices, nodes, and cloud infrastructure, facilitating the seamless operation of distributed systems. The ongoing process of network cloudification is transitioning from the use of specially designed network appliances to data center-like systems built with commodity servers and storage. In these systems, virtual network functions can be deployed as software instances and composed as service components for provisioning services. Consequently, networks are evolving from infrastructures solely dedicated to data communications into versatile platforms that support both networking and computing services. Moreover, network cloudification adopts the cloud service model for network

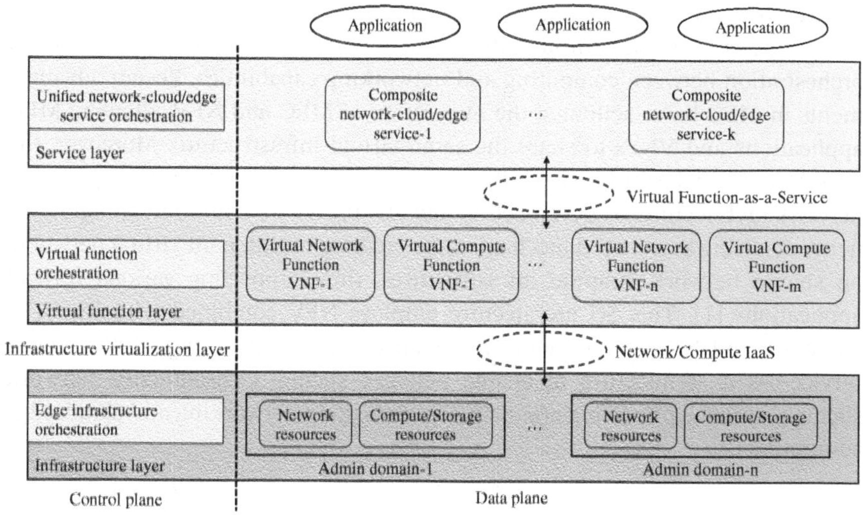

Figure 3.8 The architectural framework for edge computing-network integration.

service delivery. This approach allows data centers and edge servers, initially constructed for edge computing, to host virtual network functions. As depicted in Figure 3.8, the infrastructure layer of this framework includes several administrative domains, each managed by different infrastructure providers. These administrative domains encompass various technical domains, each specializing in a specific type of edge infrastructure, such as networking, computing, or storage. The virtualization layer abstracts these diverse computing and networking resources, offering them as virtual resources through the IaaS model. On the virtual function layer, virtual network functions (VNFs) and virtual compute functions (VCFs) are implemented using these infrastructure services. These VNFs and VCFs are then orchestrated at the service layer to form composite services, which support various applications.

3.3.1.1 The Development Process

Initially, efforts in networking concentrated on integrating computing and network infrastructures within data centers, where high-speed interconnections between servers and storage systems optimized performance. This foundational phase paved the way for further advancements in networked computing environments. A significant driving force behind network cloudification is the ETSI ISG network function virtualization (NFV) [52]. In the NFV architecture, both network and compute infrastructures are abstracted by a common virtualization layer, enabling the utilization of virtual resources to realize VNFs [8]. The Management and Orchestration (MANO) component handles the service and resource management as well as orchestration. The true potential of edge computing is harnessed through its interaction with networking, especially the orchestration between computing and networking capabilities. Recent advancements in MEC have facilitated the integration of MEC and NFV, allowing MEC applications and VNFs to share the same virtual infrastructure. Moreover, the NFV MANO can be utilized for MEC management and orchestration [15]. The 5G network architecture, as developed by the 3GPP/5G-PPP community, highlights network slicing as a crucial mechanism for creating multitenant virtual networks on shared network-compute infrastructures, thus supporting various vertical applications [1]. This 5G architecture employs NFV combined with the SDN paradigm and leverages virtualization, softwarization, programmability, and a service-based architecture to enable network slicing. Consequently, network slicing presents a promising approach for converging networking and cloud/edge computing [9].

3.3.2 Edge Computing-Network Architectures

The architecture of edge computing networks is fundamental to achieving efficient, scalable, and resilient systems. This subsection explores various architectural designs, examining their characteristics and implications for edge computing. Understanding these architectures is crucial for developing systems that can effectively leverage the benefits of edge computing while addressing the challenges of distributed networking.

3.3.2.1 Edge Network Design

Edge network design involves creating architectures that optimize the deployment and connectivity of edge nodes. Common designs include hierarchical models [17, 42], where edge nodes are organized in tiers based on their proximity to data sources and end-users, and mesh networks, where nodes are interconnected in a nonhierarchical manner, providing multiple pathways for data transmission. Hierarchical designs often facilitate easier management and scalability, while mesh networks enhance redundancy and fault tolerance. The choice of design impacts the performance, reliability, and complexity of the network, influencing factors such as latency, bandwidth utilization, and system robustness.

3.3.2.2 Distributed Networking

Distributed networking in edge computing spreads processing and data storage across multiple nodes rather than relying on a centralized infrastructure [21]. This approach enhances scalability and resilience by distributing workloads, reducing single points of failure, and allowing for localized data processing.

Distributed networks are essential for applications requiring real-time processing and low latency [36], such as IoT, smart cities, and industrial automation. By decentralizing computing tasks, these architectures can adapt to varying loads and improve the overall efficiency of the system.

3.3.2.3 Top-Down, Service-Enabled Convergence Architectures

Top-down, service-enabled convergence architectures integrate various layers of the computing stack, from hardware to applications, into a unified framework [33]. This approach allows for seamless service delivery and resource management across the edge and cloud environments. By converging networking, computing, and storage resources, these architectures support the dynamic allocation of resources based on service requirements. They facilitate the deployment of complex services that can span multiple edge nodes and cloud resources, ensuring that computing power, storage, and networking are efficiently utilized to meet application demands.

3.3.2.4 Computing Power-Aware Architectures

Computing power-aware architectures are designed to optimize the use of computational resources based on the specific needs of applications and the capabilities of edge nodes. These architectures dynamically adjust the distribution of workloads according to the available computing power, energy efficiency, and processing requirements [22, 24, 41, 56]. By being aware of the computing power at each node, these architectures can enhance performance, reduce energy consumption, and ensure that tasks are allocated to the most suitable resources. This approach is particularly important for applications with varying computational demands, such as AI-driven analytics, real-time video processing, and complex simulations.

3.3.3 Current Progress and Future Trend

In recent years, substantial progress has been made in the integration of edge computing with networking technologies. Key developments include:

- **Enhanced edge network infrastructure:** Deployment of 5G networks has dramatically improved the connectivity and bandwidth available for edge computing [16]. This infrastructure supports higher data rates, reduced latency, and greater device density, which are critical for applications such as autonomous vehicles, AR, and IoT.
- **Advanced distributed networking solutions:** Innovations in distributed networking, such as software-defined networking (SDN) [3], named-data networking (NDN) [32], and NFV, have enabled more flexible and dynamic

network configurations. These technologies allow for better resource allocation and management, ensuring that edge nodes can efficiently handle diverse workloads.

- **Edge AI integration:** The integration of AI at the edge has progressed significantly [47]. Edge AI models can now perform complex data processing and decision-making locally, reducing the need for constant data transmission to the cloud. This advancement is crucial for applications requiring real-time insights and actions, such as predictive maintenance and intelligent surveillance.
- **Improved security mechanisms:** With the increasing deployment of edge devices, security has become a paramount concern. Current progress includes the development of robust security frameworks that protect data integrity, confidentiality, and availability at the edge [18]. Techniques such as secure boot, trusted execution environments, and edge-based encryption are being widely adopted.

Looking ahead, several trends are expected to drive the future of edge computing and networking:

- **Edge-to-edge and edge-to-multi-cloud integration:** The future will see deeper integration between edge networks, enabling seamless edge-to-edge communication and collaboration. Additionally, edge-to-multi-cloud architectures will emerge, allowing edge nodes to interact with multiple cloud providers, optimizing resource use and ensuring redundancy.
- **AI-driven network management:** The application of AI and machine learning in network management will become more prevalent. AI-driven approaches will enable predictive maintenance, automated troubleshooting, and intelligent resource allocation, enhancing the overall efficiency and reliability of edge networks.
- **Expansion of computing power-aware architectures:** Future architectures will increasingly incorporate computing power-awareness, dynamically adjusting to the available resources and the specific needs of applications. This trend will lead to more efficient and energy-conscious edge computing systems.
- **Development of unified edge standards:** As edge computing continues to grow, the development of unified standards and protocols will become crucial. Standardization efforts will facilitate interoperability between different edge devices and platforms, fostering a more cohesive and scalable edge ecosystem.
- **Growth of edge ecosystems and partnerships:** Collaboration between technology providers, enterprises, and industries will drive the growth of edge ecosystems. Partnerships will enable the development of specialized edge solutions tailored to specific industry needs, accelerating the adoption of edge computing across various sectors.

3.4 Summary and Practice

3.4.1 Summary

This chapter initially delineates the constituents of the edge computing architectural framework, classifying them based on the vantage point of their concrete deployment into three distinct categories: on-premises edge, network edge, and data center edge. By fostering synergy with cloud and other peripheral devices, the edge computing architecture demonstrates robust competencies in realms such as data processing, cache and storage, communication, and CDNs.

Subsequently, this chapter articulates the definition of the edge computing paradigm and elucidates four distinct collaborative modalities intrinsic to edge computing. These encompass edge-to-edge collaboration, which facilitates interaction between disparate edge nodes; edge-to-device collaboration, which integrates edge capabilities with end-user devices; edge-to-cloud collaboration, which harnesses the power of cloud computing to complement edge processing; and cloud-edge-device collaboration, which creates a cohesive and integrated ecosystem among clouds, edges, and devices.

Finally, this chapter discusses the integration of edge computing with the network. It outlines how network integration is essential for seamless data processing and communication across distributed environments. By combining the two and developing them together, selecting the appropriate model can enhance the overall efficiency and effectiveness of the system.

3.4.2 Practice Questions

1) Discuss the different grades/layers of edge infrastructure. How do these layers contribute to the overall efficiency and scalability of an edge computing system?
2) Compare and contrast mobile edge computing (MEC) and multi-access edge computing (MEC).
3) Explain the role of gateways in an edge computing architecture.
4) What are collaborative edge computing models, and in what scenarios are they most beneficial?
5) What factors should be considered when choosing the right edge computing model for a specific application? Provide examples to support your explanation.

3.4.3 Course Projects

1) Set up a small edge computing network and demonstrate communication between edge devices and servers.

2) Design a multilayer edge computing architecture tailored to a specific application, such as smart cities, autonomous vehicles, or smart homes, and explain how each layer contributes to the overall architecture and meets the application's requirements.
3) Build a demo system for cloud-edge-device collaboration, with the functions of device-side data collection, edge-side preprocessing data, and cloud batch data processing.
4) Implement a system for managing and monitoring edge devices using KubeEdge.
5) Evaluate a few popular Kubernetes performances and discuss their advantages/disadvantages. An example Github repo: https://github.com/hkoziolek/lightweight-k8s-benchmarking.

Chapter 3 Suggested Papers

1 Pedro Cruz, Nadjib Achir, and Aline Carneiro Viana. "On the edge of the deployment: A survey on multi-access edge computing". In: *ACM Computing Surveys* 55. 5 (2022), pp. 1–34.

2 Yun Chao Hu et al. "Mobile edge computing—A key technology towards 5G". In: *ETSI White Paper* 11. 11 (2015), pp. 1–16.

3 Heiko Koziolek and Nafise Eskandani. "Lightweight Kubernetes distributions: A performance comparison of MicroK8s, k3s, k0s, and MicroShift". In: *Proceedings of the 2023 ACM/SPEC International Conference on Performance Engineering*. 2023, pp. 17–29.

4 Fang Liu et al. "A survey on edge computing systems and tools". In: *Proceedings of the IEEE* 107. 8 (2019), pp. 1537–1562.

References

1 5G PPP Architecture Working Group. *View on 5G Architecture: Version 3.0*. Tech. rep. 5G PPP Association, 2019.

2 Mohammad Babar et al. "Cloudlet computing: Recent advances, taxonomy, and challenges". In: *IEEE Access* 9 (2021), pp. 29609–29622.

3 Ahmet Cihat Baktir, Atay Ozgovde, and Cem Ersoy. "How can edge computing benefit from software-defined networking: A survey, use cases, and future directions". In: *IEEE Communications Surveys & Tutorials* 19. 4 (2017), pp. 2359–2391.

4 Sebastian Böhm and Guido Wirtz. "Profiling Lightweight Container Platforms: MicroK8s and K3s in Comparison to Kubernetes". In: *ZEUS*. 2021, pp. 65–73.

5 Canonical Ltd. *MicroK8s - Zero-ops Kubernetes for developers, edge, and IoT.* https://microk8s.io/. Accessed: 2024-08-29.

6 Xiaofeng Cao et al. "Edge federation: Towards an integrated service provisioning model". In: *IEEE/ACM Transactions on Networking* 28. 3 (2020), pp. 1116–1129.

7 Pedro Cruz, Nadjib Achir, and Aline Carneiro Viana. "On the edge of the deployment: A survey on multi-access edge computing". In: *ACM Computing Surveys* 55. 5 (2022), pp. 1–34.

8 Richard Cziva and Dimitrios P Pezaros. "Container network functions: Bringing NFV to the network edge". In: *IEEE Communications Magazine* 55. 6 (2017), pp. 24–31.

9 Antonio De la Oliva et al. "5G-TRANSFORMER: Slicing and orchestrating transport networks for industry verticals". In: *IEEE Communications Magazine* 56. 8 (2018), pp. 78–84.

10 Qiang Duan and Shangguang Wang. "Network cloudification enabling network-cloud/fog service unification: State of the art and challenges". In: *2019 IEEE World Congress on Services (SERVICES).* Vol. 2642. IEEE. 2019, pp. 153–159.

11 Qiang Duan, Shangguang Wang, and Nirwan Ansari. "Convergence of networking and cloud/edge computing: Status, challenges, and opportunities". In: *IEEE Network* 34. 6 (2020), pp. 148–155.

12 Melike Erol-Kantarci and Sukhmani Sukhmani. "Caching and computing at the edge for mobile augmented reality and virtual reality (AR/VR) in 5G". In: *Ad Hoc Networks: 9th International Conference, AdHocNets 2017, Niagara Falls, ON, Canada, September 28–29, 2017, Proceedings.* Springer. 2018, pp. 169–177.

13 Matthew Finnegan. *Boeing 787s to create half a terabyte of data per flight, says Virgin Atlantic.* https://www.computerworlduk.com/data/boeing-787s-create-half-terabyte-of-data-per-flight-says-virgin-atlantic-3433595 (2013).

14 General Data Protection Regulation GDPR. "General data protection regulation". In: *Regulation (EU) 2016/679 of the European Parliament and of the Council of 27 April 2016 on the protection of natural persons with regard to the processing of personal data and on the free movement of such data, and repealing Directive 95/46/EC* (2016).

15 Kai Han et al. "Application-driven end-to-end slicing: When wireless network virtualization orchestrates with NFV-based mobile edge computing". In: *IEEE Access* 6 (2018), pp. 26567–26577.

16 Najmul Hassan, Kok-Lim Alvin Yau, and Celimuge Wu. "Edge computing in 5G: A review". In: *IEEE Access* 7 (2019), pp. 127276–127289.

17 Qiang He et al. "Pyramid: Enabling hierarchical neural networks with edge computing". In: *Proceedings of the ACM Web Conference 2022.* 2022, pp. 1860–1870.

18 Size Hou and Xin Huang. "Use of machine learning in detecting network security of edge computing system". In: *2019 IEEE 4th International Conference on Big Data Analytics (ICBDA)*. IEEE. 2019, pp. 252–256.

19 Yun Chao Hu et al. "Mobile edge computing—A key technology towards 5G". In: *ETSI White Paper* 11. 11 (2015), pp. 1–16.

20 Han Hu et al. "Intelligent resource allocation for edge-cloud collaborative networks: A hybrid DDPG-D3QN approach". In: *IEEE Transactions on Vehicular Technology* 72. 8 (2023), pp. 10696–10709.

21 Cheng-Fu Huang, Ding-Hsiang Huang, and Yi-Kuei Lin. "Network reliability evaluation for a distributed network with edge computing". In: *Computers & Industrial Engineering* 147 (2020), p. 106492.

22 Yi-Wen Hung et al. "Dynamic workload allocation for edge computing". In: *IEEE Transactions on Very Large Scale Integration (VLSI) Systems* 29. 3 (2021), pp. 519–529.

23 David Jensen. *Beginning Azure IoT Edge computing: Extending the cloud to the intelligent edge*. Apress, 2019.

24 Congfeng Jiang et al. "Energy aware edge computing: A survey". In: *Computer Communications* 151 (2020), pp. 556–580.

25 Agus Kurniawan. *Learning AWS IoT: Effectively manage connected devices on the AWS cloud using services such as AWS Greengrass, AWS button, predictive analytics and machine learning*. Packt Publishing Ltd, 2018.

26 Li Li et al. "A review of applications in federated learning". In: *Computers & Industrial Engineering* 149 (2020), p. 106854.

27 Chunlin Li et al. "Smart contract-based decentralized data sharing and content delivery for intelligent connected vehicles in edge computing". In: *IEEE Transactions on Intelligent Transportation Systems* 25. 10 (2024), pp. 14535–14545.

28 Ze Liu et al. "Swin transformer: Hierarchical vision transformer using shifted windows". In: *Proceedings of the IEEE/CVF International Conference on Computer Vision*. 2021, pp. 10012–10022.

29 Pavel Mach and Zdenek Becvar. "Mobile edge computing: A survey on architecture and computation offloading". In: *IEEE Communications Surveys & Tutorials* 19. 3 (2017), pp. 1628–1656.

30 Satyanarayanan Mahadev et al. "Cloudlets: At the leading edge of mobile-cloud convergence". In: *2014 6th International Conference on Mobile Computing, Applications and Services (MobiCASE))*. IEEE. 2014, pp. 1–9.

31 Yuyi Mao et al. "A survey on mobile edge computing: The communication perspective". In: *IEEE Communications Surveys & Tutorials* 19. 4 (2017), pp. 2322–2358.

32 Named Data Networking. "Named Data Networking". In: *Named Data Networking* 20 (2012).

33 Kai Peng et al. "A survey on mobile edge computing: Focusing on service adoption and provision". In: *Wireless Communications and Mobile Computing* 2018. 1 (2018), p. 8267838.

34 Kai Peng et al. "End-edge-cloud collaborative computation offloading for multiple mobile users in heterogeneous edge-server environment". In: *Wireless Networks* 30 (2020), pp. 3495–3506.

35 Rancher. *Lightweight Certified Kubernetes Distribution K3s.* https://www.rancher.com/products/k3s. Accessed: 2024-08-29.

36 Pradip Kumar Sharma et al. "SoftEdgeNet: SDN based energy-efficient distributed network architecture for edge computing". In: *IEEE Communications Magazine* 56. 12 (2018), pp. 104–111.

37 Yuanming Shi et al. "Communication-efficient edge AI: Algorithms and systems". In: *IEEE Communications Surveys & Tutorials* 22. 4 (2020), pp. 2167–2191.

38 Ashish Singh and Kakali Chatterjee. "Securing smart healthcare system with edge computing". In: *Computers & Security* 108 (2021), p. 102353.

39 Ion Stoica and Scott Shenker. "From cloud computing to sky computing". In: *Proceedings of the Workshop on Hot Topics in Operating Systems.* 2021, pp. 26–32.

40 Yevgeniy Sverdlik. *Here's How Much Energy All US Data Centers Consume.* https://www.datacenterknowledge.com/epower-supply/here-s-how-much-energy-all-us-data-centers-consume (2016).

41 Minh-Tuan Thai et al. "Workload and capacity optimization for cloud-edge computing systems with vertical and horizontal offloading". In: *IEEE Transactions on Network and Service Management* 17. 1 (2019), pp. 227–238.

42 Liang Tong, Yong Li, and Wei Gao. "A hierarchical edge cloud architecture for mobile computing". In: *IEEE INFOCOM 2016-The 35th Annual IEEE International Conference on Computer Communications.* IEEE. 2016, pp. 1–9.

43 Ashish Vaswani et al. "Attention is all you need". In: *Advances in Neural Information Processing Systems* 30 (2017).

44 Yimeng Wang et al. "IndustEdge: A time-sensitive networking enabled edge-cloud collaborative intelligent platform for smart industry". In: *IEEE Transactions on Industrial Informatics* 18. 4 (2021), pp. 2386–2398.

45 Rui Wang et al. "Survey of edge–edge collaborative training for edge intelligence". In: *Chinese Journal of Engineering* 45. 8 (2023), pp. 1400–1416.

46 Yingchao Wang et al. "End-edge-cloud collaborative computing for deep learning: A comprehensive survey". In: *IEEE Communications Surveys & Tutorials* 26. 4 (2024), pp. 2647–2683.

47 Dingzhu Wen et al. "Task-oriented sensing, computation, and communication integration for multi-device edge AI". In: *IEEE Transactions on Wireless Communications* 23. 3 (2023), pp. 2486–2502.

48 Dapeng Wu et al. "Edge-cloud collaboration enabled video service enhancement: A hybrid human-artificial intelligence scheme". In: *IEEE Transactions on Multimedia* 23 (2021), pp. 2208–2221.

49 Ying Xiong et al. "Extend cloud to edge with Kubeedge". In: *2018 IEEE/ACM Symposium On Edge Computing (SEC)*. IEEE. 2018, pp. 373–377.

50 Jiangchao Yao et al. "Edge-cloud polarization and collaboration: A comprehensive survey for AI". In: *IEEE Transactions on Knowledge and Data Engineering* 35. 7 (2022), pp. 6866–6886.

51 Hikmat Yar et al. "Towards smart home automation using IoT-enabled edge-computing paradigm". In: *Sensors* 21. 14 (2021), p. 4932.

52 Bo Yi et al. "A comprehensive survey of network function virtualization". In: *Computer Networks* 133 (2018), pp. 212–262.

53 Chen Zhang et al. "A survey on federated learning". In: *Knowledge-Based Systems* 216 (2021), p. 106775.

54 Yongli Zhao et al. "Edge computing and networking: A survey on infrastructures and applications". In: *IEEE Access* 7 (2019), pp. 101213–101230.

55 Yuhan Zhao et al. "A survey on caching in mobile edge computing". In: *Wireless Communications and Mobile Computing* 2021. 1 (2021), p. 5565648.

56 Fuhui Zhou and Rose Qingyang Hu. "Computation efficiency maximization in wireless-powered mobile edge computing networks". In: *IEEE Transactions on Wireless Communications* 19. 5 (2020), pp. 3170–3184.

4

Toward Edge Intelligence*

In this chapter, we transition from the foundational concepts of edge computing to the emerging field of edge intelligence (EI). As we explore this exciting frontier, we will examine how artificial intelligence (AI) is integrated into edge computing systems to benefit humans and society.

4.1 What Is Edge Intelligence?

With the burgeoning growth of the Internet of Everything, the amount of data generated by edge increases dramatically, resulting in higher network bandwidth requirements. Meanwhile, the emergence of novel applications calls for lower latency of the network. Based on these two main requirements, edge computing arises, which refers to processing the data at the edge of the network. Edge computing guarantees quality of service when dealing with a massive amount of data for cloud computing [99].

At the same time, AI applications based on machine learning (especially deep learning algorithms) are fueled by advances in models, processing power, and big data. Nowadays, applications are built as a central attribute, and users are beginning to expect near-human interaction with the appliances they use. For example, mobile phone applications, such as those related to face recognition and speech translation, have a high requirement to run online or offline.

As shown in Figure 4.1 pushed by edge computing techniques and pulled by AI applications, EI has been pushed to the horizon. The development of edge computing techniques, including powerful Internet of Things (IoT) data, edge devices, storage, wireless communication, and security and privacy, make it possible to run AI algorithms on the edge. AI applications, including connected

*This chapter is contributed by Shaibal Saha, Qiren Wang, Lichen Xia, and Yongtao Yao.

Edge Computing: Systems and Applications, First Edition. Lanyu Xu and Weisong Shi.

Companion Website: www.wiley.com/go/xu/computing

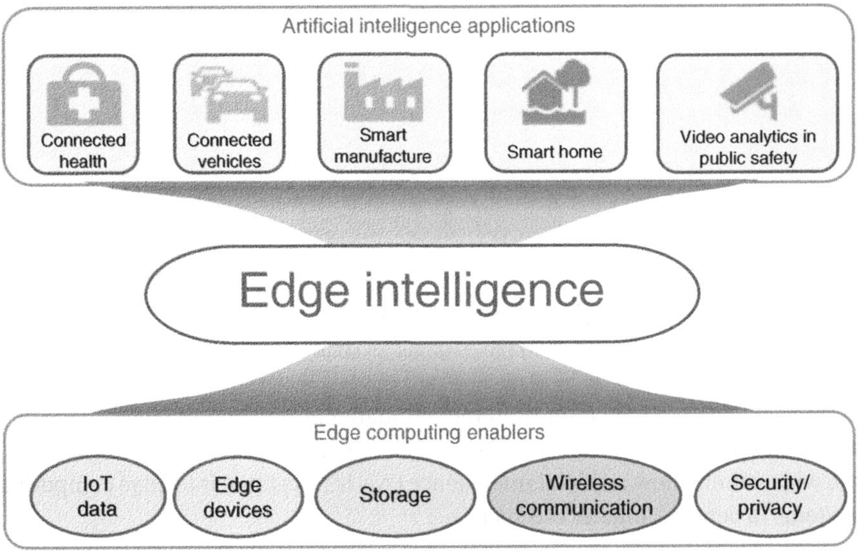

Figure 4.1 Motivation of edge intelligence.

health, connected vehicles, smart manufacturing, smart home, and video analytics, require running on edge. In the EI scenario, advanced AI models based on machine learning algorithms will be optimized to run on the edge. The edge will be capable of dealing with video frames, natural speech information, time series data, and unstructured data generated by cameras, microphones, and other sensors without uploading data to the cloud and waiting for the response.

4.1.1 Formal Definition

International Electrotechnical Commission (IEC) defines EI as the process by which data are acquired, stored, and processed with machine learning algorithms on the network edge. It believes that several industries of information technology and operational technology are moving closer to the edge of the network so that aspects such as real-time networks, security capabilities, and personalized/customized connectivity are addressed [49]. In 2018, [90] discussed the challenges and the opportunities that EI created by presenting a use-case showing that the careful design of the convolutional neural networks (CNNs) for object detection would lead to real-time performance on embedded edge devices. [115] leveraged EI for activity recognition in smart homes from multiple perspectives, including architecture, algorithm, and system.

In this book, we refer to the definition from [117]: EI is defined as **the capability to enable edges to execute artificial intelligence algorithms**.

The diversity of edge hardware results in differences in AI models or algorithms they carry; that is, edges have different EI capabilities. The capability here is defined as a four-element tuple < *Accuracy, Latency, Energy, Memory footprint* > which is abbreviated as ALEM. *Accuracy* is the internal attribute of AI algorithms. In practice, the definition of *Accuracy* depends on specific applications; for example, it is measured by mean average precision in object detection tasks and measured by the Bilingual Evaluation Understudy score metric in machine translation tasks. To execute the AI tasks on the edge, some algorithms are optimized by compressing the size of the model, quantizing the weight, and other methods that will decrease accuracy. Better EI capability means that the edge is able to employ the algorithms with greater Accuracy. *Latency* represents the inference time when the trained model is run on the edge. To measure *Latency*, the average latency of multiple inference tasks can be calculated. When running the same models, the *Latency* measures the level of edge performance. *Energy* refers to the increased power consumption of the hardware when executing the inference task. *Memory footprint* is the memory usage when running the AI model. *Energy* and *Memory footprint* indicate the computing resource requirements of the algorithms.

There are two types of collaboration for EI: cloud-edge and edge-edge collaboration. In the cloud-edge scenario, the models are usually trained on the cloud and then downloaded to the edge, which executes the inference task. Sometimes, edges will retrain the model by transfer learning based on the data they generate. The retrained models will be uploaded to the cloud and combined into a general and global model. In addition, researchers have focused on distributed deep learning models over the cloud and on the edge. For example, DDNN [103] is a distributed deep neural network architecture across the cloud and edge. Edge–edge collaboration has two aspects. First, multiple edges work collaboratively to accomplish a compute-intensive task. For example, several edges will be distributed when training a large deep-learning network. The task will be allocated according to the computing power. Second, multiple edges work together to accomplish a task with different divisions based on different environments. For example, in smart home environments, a smartphone predicts when a user is approaching home, triggering the smart thermostat to set the suitable temperature for the user. Individually, every task is particularly difficult, but coordination within the edge makes it easy.

As shown in Figure 4.2, the data generated by the edge come from different sources, such as cars, drones, smart homes, etc., and can be used in three different ways:

- First is uploading the data to the cloud and training based on the multisource data. When the model training is completed, the cloud will do the inference

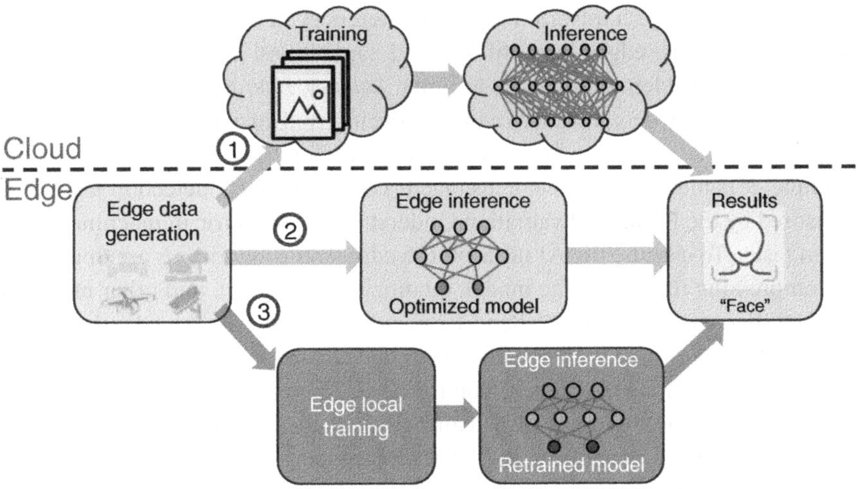

Figure 4.2 Dataflow of edge intelligence.

based on the edge data and send the result to the edge. This data flow is widely used in traditional machine intelligence.

- Second is executing the inference on the edge directly. The data generated by the edge will be the input of the edge model downloaded from the cloud. The edge will do the inference based on the input and output of the results. This is the current EI data flow.
- Third is training on the edge locally. The data will be used to retrain the model on the edge by taking advantage of transfer learning. After retraining, the edge will build a personalized model that has better performance for the data generated on the edge. This will be the future data flow of EI.

EI involves a great deal of knowledge and technology, such as the design of AI algorithms, software and systems, computing architecture, sensor networks, and so on. Figure 4.3 shows the overview of EI. To support EI, many techniques have been developed, called EI techniques, which include algorithms, software, and hardware. Representative EI techniques will be introduced in the remainder of the chapter.

4.2 Hardware and Software Support

EI relies heavily on hardware and software advancements to function effectively and bring AI capabilities to the edge of the network. In this section, we'll explore the specific hardware components, such as the specialized processors

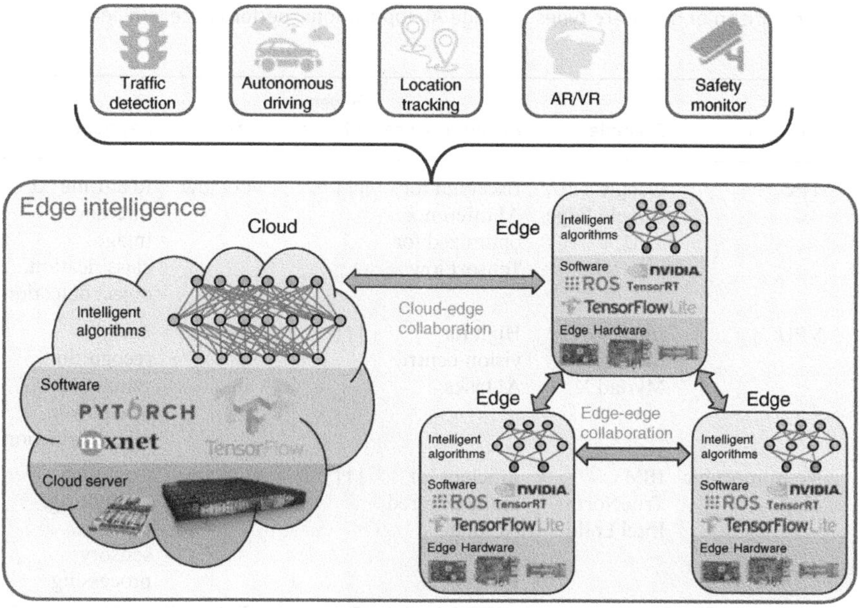

Figure 4.3 Edge intelligence.

and accelerators, that are essential for enabling AI at the edge. We'll also discuss the software frameworks and platforms that support the development and deployment of EI.

4.2.1 Hardware

Unlike traditional cloud centers, edge devices are often supplied with constrained computation and power resources and must accommodate different end-user requirements. In Table 4.1, we list typical hardware for deploying EI, each serving different functions based on their strengths.

4.2.1.1 Application-Specific Integrated Circuit (ASIC)

An application-specific integrated circuit (ASIC) is an integrated circuit that is custom-designed for a particular task or application. ShiDianNao [32] first proposed that the AI processor should be deployed next to the camera sensors. The processor accesses the image data directly from the sensor instead of dynamic random access memory (DRAM), which reduces the power consumption of the sensor data loading and storing. ShiDianNao is 60 times more energy efficient and 30 times faster than the previous state-of-the-art AI hardware, so it will be suitable for EI applications related to computer vision. Efficient inference engine

Table 4.1 Comparison of hardware types for edge AI applications: performance, efficiency, and suitability.

Hardware type	Sub-type	Examples	AI performance	Power efficiency	Latency	Key applications
ASIC	TPU	Google TPU, Google Edge TPU	Excellent for AI inference, optimized for TensorFlow	↑↑↑	Very low	Real-time AI inference, image classification, object detection
	VPU	Intel Movidius Myriad X	High for vision-centric AI tasks	↑↑↑	Low	Facial recognition, visual SLAM, low-power object detection
	Neuromorphic chip	IBM TrueNorth, Intel Loihi	Efficient for brain-inspired AI models	↑↑↑↑	Very low	Cognitive computing, robotics, sensory processing
FPGA		Xilinx UltraScale+, Intel Stratix	High, customizable for specific AI workloads	↑↑	Low to moderate	Custom AI models, adaptive AI tasks
GPU		Jetson Nano, AGX Xavier	High, particularly in parallel processing tasks	↑	Medium	Deep learning, autonomous systems, AI research

↑↑↑↑: extremely high; ↑↑↑: very high; ↑↑: moderate; ↑: high.

(EIE) [40] is an efficient hardware design for compressed deep neural networks (DNN) inference. Using multiple methods to improve energy efficiency, such as exploiting the sparsity of DNN and sharing the weights of DNN, it is deployed on mobile devices to process some embedded EI applications. In industry, many leaders have published some dedicated hardware modules to accelerate EI applications; for example, IBM TrueNorth [83] and Intel Loihi [26] are both neuromorphic processors.

Google Cloud introduced Edge TPU (Figure 4.4) customized for ML inference on edge devices [35]. Microsoft's Azure Sphere is a security-focused microcontroller that incorporates an ASIC designed to provide hardware-based security features for IoT devices, ensuring robust protection against various security threats [82]. AWS Snowball [3] and Snowcone [4] are portable, rugged, and

Figure 4.4 TPU.

secure edge computing devices that collect, process, and move data to AWS from disconnected environments.

Another significant development is Intel's Movidius Myriad X. This vision processing unit (VPU) (Figure 4.5) integrates 16 SHAVE (Streaming Hybrid Architecture Vector Engine) cores and a dedicated neural compute engine for deep learning inference. The Myriad X delivers over 1 TOPS of computational performance, enabling sophisticated AI applications such as object detection, facial recognition, and autonomous navigation on edge devices.

NVIDIA has also made significant strides with its Jetson Nano, a small yet powerful AI computer that delivers 472 GFLOPs of computational power. It supports multiple neural networks in parallel for applications such as image classification, object detection, and speech processing. The Jetson Nano (Figure 4.6)

Figure 4.5 VPU.

Figure 4.6 Jetson Nano.

Figure 4.7 TrueNorth chip.

is designed to run on just 5–10 watts of power, making it suitable for embedded IoT applications and autonomous machines (Figure 4.9).

IBM's TrueNorth neuromorphic chip represents a different approach to ASIC design (Figure 4.7). Inspired by the human brain, TrueNorth is designed to process information in a highly parallel and efficient manner, mimicking the brain's neural network architecture. This chip excels in applications requiring real-time pattern recognition and sensory processing, such as robotics and cognitive computing.

Apple also proposed the Apple Neural Engine (ANE), which is designed to accelerate machine learning tasks within Apple's suite of mobile devices. Introduced as part of the A11 Bionic chip and continually evolving in subsequent models, the ANE is a specialized ASIC designed to accelerate neural network operations. The ANE enhances performance, reduces latency, and maintains user privacy by enabling local processing of tasks such as voice recognition, facial recognition, and augmented reality.

4.2.1.2 Field Programmable Gate Arrays (FPGAs) and Graphics Processing Units (GPUs)

Several studies have deployed general-purpose field programmable gate arrays (FPGAs, as shown in Figure 4.8) or graphics processing units (GPUs) for EI application scenarios, such as speech recognition, image classification, and

Figure 4.8 FPGA.

natural language processing (NLP). Both Intel and Xilinx are pioneer companies in the field of FPGA research. For example, Xilinx has introduced a series of specialized adaptable accelerator solutions catering to different AI tasks. The Alveo accelerator cards are designed for high-performance data center applications, providing powerful acceleration for workloads such as machine learning inference (especially deep learning) and video processing. On the other hand, Xilinx also launched multiprocessor system-on-chip (MPSoC) devices, which are ideal for edge computing applications. These MPSoC devices combine arm-based processors and programmable logic gates, making them highly suitable for AI inference tasks at the edge where power efficiency and real-time processing are crucial. The series of MPSoC FPGAs are highly used for deep learning models due to its robust software ecosystem that includes support for high-level programming languages. Similarly, Intel has developed the Agilex and Stratix 10 FPGA families, which are aimed at high-performance acceleration tasks and are supported by a Quad-core ARM Cortex-A53 processor. In the current scenario, Xilinx has a mature and widely used development environment, particularly with its Vivado Design suite, which is considered user-friendly. ESE [41] used FPGAs to accelerate the LSTM model on mobile devices, which adopted the load-balance-aware pruning method to ensure high hardware utilization and the partitioned compressed LSTM model on multiple processing elements (PEs) to process the LSTM data flow in parallel. The implementation of ESE on Xilinx FPGA achieved higher energy efficiency compared with the CPU and GPU. Biookaghazadeh et al. used a specific EI workload to evaluate the performance of FPGA and GPU on edge devices. They compared some metrics, such as data throughput and energy efficiency, between the FPGA and GPU. The evaluation results showed that the FPGA is more suitable for EI application scenarios [13].

In industry, NVIDIA published the Jetson AGX Xavier module (Figure 4.9) [87], which is equipped with a 512-core Volta GPU and an 8-core ARM 64-bit CPU. It supports CUDA and TensorRT libraries to accelerate AI applications in several

Figure 4.9 Nvidia AGX Xavier.

scenarios, such as robot systems and autonomous vehicles. In 2022, NVIDIA also introduced the Jetson Orin box [85], which provides even higher performance and efficiency and supports more advanced AI models, making it suitable for complex edge AI applications.

4.2.2 Software

While hardware forms the foundation of edge intelligence by providing the necessary computational power and specialized processing capabilities, the software enables advanced functionality, transforming raw processing power into intelligent, efficient, and scalable edge solutions.

4.2.2.1 Packages

In order to execute AI applications efficiently, many deep learning packages are specifically designed to meet the computing paradigm of AI algorithms, such as TensorFlow, Caffe, MXNet, and PyTorch. However, these packages are focused on the cloud and are not suitable for the edge. In the cloud, packages use a large-scale dataset to train deep-learning models. One of the main tasks of packages is to learn the number of weights in each model layer. They are deployed on the high-performance platforms, such as GPU, CPU, FPGA, and ASIC (TPU [54]) clusters. On the edges, due to limited resources, packages do not train models in most cases. They carry on inference tasks by leveraging the models which have been trained in the cloud. The input is small-scale real-time data and the packages are installed on heterogeneous edges, such as edge servers, mobile phones, Raspberry Pi, and laptops.

To support processing data and executing AI algorithms on the edges, some top-leading tech giants have released several edge-based deep learning packages. Compared with cloud versions, these frameworks require significantly fewer resources but behave almost the same in terms of inference. TensorFlow Lite [104] is TensorFlow's lightweight solution designed for mobile and edge devices. It leverages many optimization techniques, including optimizing the kernels for mobile apps, prefused activations, and quantized kernels to reduce the latency. ONNX (Open Neural Network Exchange) [89] runtime is another open-source performance-focused inference engine for machine learning models. It is a part and parcel of the ONNX ecosystem launched by Facebook and Microsoft. It supports models in the ONNX format, which is a cross-platform standard designed to make models portable across different hardware and software. ONNX runtime is optimized for a variety of platforms, including Jetson Nano, Xilinx FPGAs, NPU, and Nvidia AGX. Apple published CoreML [11], a deep learning package optimized for on-device performance to minimize memory footprint and power consumption. Users are allowed to integrate the trained machine learning model into Apple products, such as Siri, Camera, and QuickType. Facebook developed QNNPACK (Quantized Neural Networks PACKage) [77], which is a mobile-optimized library for high-performance neural network inference. It provides an implementation of common neural network operators on quantized 8-bit tensors. Moreover, Google developed XNNPACK [34] for highly efficient floating-point neural network inference on ARM, x86, webAssembly, and RISC-V platforms. XNNPACK serves as a low-level performance primitive aimed at accelerating high-level machine learning frameworks, including TensorFlow Lite, TensorFlow.js, PyTorch, ONNX Runtime, and MediaPipe, rather than being designed for direct usage by deep learning researchers. Microsoft introduced Azure IoT Edge [81], which enables deployment and management of cloud-native workloads (e.g., AI, Azure services, or user's own business logic) on IoT devices. Intel's OpenVINO toolkit is an open-source toolkit that accelerates AI inference with lower latency and higher throughput while maintaining accuracy [48]. It converts and optimizes models trained using popular frameworks such as TensorFlow and PyTorch, and then deploys the optimized models across a mix of Intel®hardware and environments, on-premise and on-device, in the browser, or in the cloud.

In addition to the edge-optimized packages such as TensorFlow Lite and Core ML, Vitis AI [9] by Xilinx stands out as a crucial development by developing AI applications on Xilinx's FPGA and SOC. Vitis AI leverages the hardware acceleration capabilities of Xilinx devices, offering model optimization, compiler, runtime, and quantization. Vitis AI includes a comprehensive suite of libraries and APIs that support popular machine learning frameworks like TensorFlow and PyTorch, authorizing developers to seamlessly import and optimize their pretrained models (AI model zoo) for execution on Xilinx platforms. Furthermore, Vitis AI provides an AI compiler, including high-level network description into low-level hardware instructions, and an AI profiler like Vaitrace for evaluating the efficiency of the inference. Additionally, FINN (Fast, Intuitive Neural Network compiler) [106] is part of the broader ecosystem of tools for accelerating machine learning applications on Xilinx hardware. FINN was developed by the AI Lab of AMD Research & Advanced Development. FINN delivers an end-to-end pipeline for taking a neural network from a quantized neural network and transforming it into an FPGA-compatible format, optimizing it for performance. It is not a generic DNN acceleration solution but relies on codesign and design space for quantization and parallelization tuning to optimize a solution. This compiler is still under continuous development but can be an excellent asset for FPGA research for AI solutions.

Meanwhile, cloud-based packages are also starting to support edge devices, such as MXNet [21] and TensorRT [86]. MXNet is a flexible and efficient library for deep learning. It is designed to support multiple platforms (either cloud platforms or edge ones) and execute training and inference tasks. TensorRT is a platform for high-performance deep learning inference, not training, and will be deployed on the cloud and edge platforms. In addition to the frameworks mentioned earlier, Amazon SageMaker Neo [6] is another tool for deploying machine learning models on edge devices. It optimizes the trained machine learning models for inferences according to the targeted device and then runs those on edge devices faster with no loss in accuracy. Amazon SageMaker Neo aims to minimize the time researchers spend manually tuning models to perform efficiently on hardware-constrained devices. Amazon SageMaker Neo utilizes Apache TVM, partner-provided compilers, and acceleration libraries to deliver the best available performance for a given model and hardware target. Several techniques, including weight and activation precision calibration, layer and tensor fusion, kernel autotuning, and multistream execution are used to accelerate the inference process. Zhang et al. [116] made a comprehensive performance comparison of several state-of-the-art deep learning frameworks on the edges and evaluated the latency, memory footprint, and energy of these frameworks with two popular deep learning models on different edge devices. They found that no framework could achieve the best performance in all dimensions, which indicated that there was a large space to improve the performance of AI frameworks on the edge. It is very important and urgent to develop a lightweight, efficient, and highly scalable framework to support AI applications on the edges.

4.2.2.2 Running Environment

To effectively support EI tasks, EI running environments must be customized and lightweight, ensuring deployment across various heterogeneous hardware platforms. They should manage diverse computational resources efficiently, handling typical workloads such as model inference and collaborative model training. The running environment needs to support deep learning packages, ensuring compatibility with the common frameworks and tools used in EI. In addition, it must handle real-time data processing from various sources, such as environmental sensors, cameras, and LiDAR, which often possess spatial and temporal attributes. Ensuring fault tolerance and optimal resource utilization is crucial, as edge computing environments require high reliability due to their proximity to data sources and limited computational power.

Taking the aforementioned requirements into account, some studies can be recognized as potential systems to support EI:

TinyOS: TinyOS [60] is an application-based operating system for sensor networks. The biggest challenge TinyOS has solved is to handle concurrency-intensive operations with small physical size and low power consumption [43]. TinyOS takes an event-driven design which is composed of a tiny scheduler and a components graph. The event-driven design makes TinyOS achieve great success in sensor networks. However, enabling effective computation migration is still a big challenge for TinyOS.

ROS and ROS2: Robot Operating System (ROS) [92] is recognized as a typical representative of the next generation of mobile operating systems designed to cope with the IoT. Originally designed to manage communication in heterogeneous robotic systems, ROS has evolved to be a versatile tool for edge computing applications. In ROS, the process that performs computations is called a node. For each service, the program or features are divided into several small pieces and distributed across multiple nodes, with the ROS topic defined to share messages between these nodes. This communication-based design gives ROS high reusability for robotics software development. ROS 2.0 [76] enhances its capabilities by offering Data Distribution Service (DDS) [27] for improved communication efficiency and addressing the issue of ROS's dependency on the master node. The active community and the formation of a robust ecosystem put ROS in a good position to be widely deployed for edge devices, including industrial robots [8, 51, 95], autonomous vehicles (e.g., Autoware) and unmanned aerial vehicles (e.g., PX4 Autopilot [91]). However, neither ROS nor ROS2 is fundamentally designed for resource allocation and computation migration, presenting challenges in the implementation of EI services directly with them.

AWS IoT Greengrass: AWS IoT Greengrass [5] is an IoT open-source edge runtime and cloud service developed by Amazon that helps users perform data management and deploy AI applications on millions of devices in homes, factories, vehicles, and businesses.

OpenVDAP: OpenVDAP [114] is an edge-based data analysis platform for Connected and Autonomous Vehicles (CAVs). OpenVDAP is a full-stack platform that contains Driving Data Integrator (DDI), Vehicle Computing Units (VCU), edge-based vehicle operating system (EdgeOSv), and libraries for vehicular data analysis (libvdap). Inside OpenVDAP, VCU supports EI by allocating hardware resources according to an application and libvdap supports EI by providing multiversions of models to accelerate model inference.

EdgeOS_H: EdgeOS_H [17] is an edge operating system designed for smart home applications. It is deployed at the edge of the home network and connects smart home devices and applications through a three-layer functional abstraction. This system addresses the diverse computing requirements of various edge hardware devices in smart homes. It emphasizes flexibility, scalability, isolation, and reliability in service management. The Phi-Stack architecture, which it incorporates, further enhances these capabilities. Additionally, the lightweight REST engine and Lua interpreter in PhiOS enable the execution of computing tasks on edge devices within the home network.

Amazon sageMaker Edge: Amazon sageMaker Edge [7] is a broader Amazon sageMaker suite that deploys, manages, and runs machine learning models on edge devices. SageMaker edge compiler compiles the trained model into an executable format that applies performance optimizations and can make the model run up to 25× faster on the targeted hardware. The edge manager also provides a dashboard to understand the performance of models on each device along with overall fleet health. SageMaker edge supports most of the machine learning frameworks such as MXNet, ONNX, and Keras.

PYNQ: PYNQ (python productivity for Zynq) [112] is an open-source python-based runtime environment designed for Xilinx Zynq SoCs, which can integrate both FPGA and ARM processors. Pynq employs the FPGA for hardware-accelerated tasks without writing traditional hardware description languages like Verilog or VHDL. It also provides a Jupyter Notebook interface for interacting with FPGA hardware, enabling high-level Python code. It includes libraries and APIs that abstract the complexity of FPGA programming, enabling real-time data processing, machine learning inference, and other computational tasks instantly on edge devices. Moreover, PYNQ enables rapid prototyping to deploy the model on FPGA, which means it helps researchers who do not have deep expertise in hardware design languages. Overall, PYNQ makes FPGA technology more accessible, enabling to harness the power of programmable hardware.

4.2.3 Container

The variety of packages and runtime environments, on the one hand, enables EI to handle diverse tasks, but on the other hand, it complicates the deployment of EI, considering it is typically a distributed system involving heterogeneous devices. Containers, which facilitate the deployment and management of EI, are therefore essential.

A container is a software component that includes all the necessary elements for running specific applications in any environment. It packages the runtime code, the execution environment, required packages and libraries, and any other dependencies of the applications into executable software units, making them portable across a variety of computing environments.

Figure 4.10 illustrates how containers help deploy and execute different applications. Since containers share the OS kernel, they do not require a full OS for each application, which keeps the size of container files small and makes container execution resource-efficient. Furthermore, because all dependencies are bundled with the application in the container, migrating applications (from the development environment to the production environment or between different computing platforms) requires minimal changes.

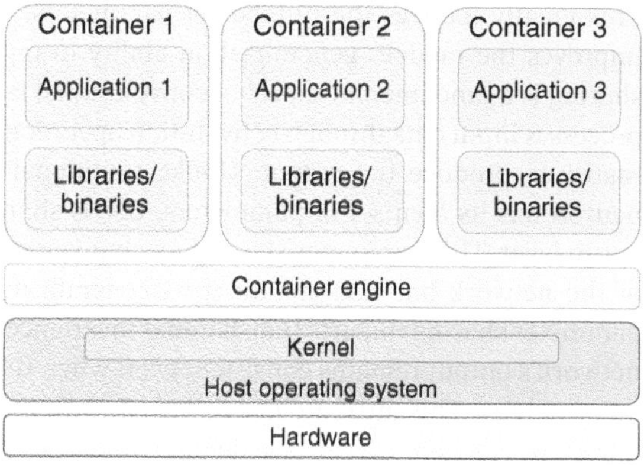

Figure 4.10 Container.

A container engine, also known as a container runtime, is a software program that creates and manages containers based on container images. It acts as an intermediary between the containers and the operating system, providing and managing resources for the application. Container engines can also process user requests, such as command line options and image pulls. Some widely used container engines include: Docker [80], RedHat RKT [94], Canonical LXD [16], and Google Kubernetes Engine [36].

4.3 Technologies Enabling Edge Intelligence

Now, we turn our focus to the cutting-edge technologies that enable these systems to operate efficiently in resource-constrained environments.

4.3.1 Compression Techniques

To enable deep learning models in edge environments, where computational resources are limited, a range of compression techniques are employed. Compression technique accelerates the speed of model inference, and are roughly categorized into four groups: parameter sharing and pruning methods, quantization, low-rank approximation methods, and knowledge distillation methods [23, 38].

4.3.1.1 Parameter Sharing and Pruning

Parameter sharing and pruning control the capacity and storage cost by reducing the number of parameters that are not sensitive to performance. In the parameter-sharing mechanism, each neuron in the neural network does not independently have a weight matrix. Instead, it shares the same weight matrix.

This greatly reduces the number of parameters, decreases the complexity, and improves the model's generalization ability to input data. In CNNs, parameter sharing is a fundamental aspect of convolutional layers. Each convolutional layer processes input data through convolution operations, utilizing a shared parameter matrix to produce the output. Unlike traditional neural networks where each neuron has its own set of parameters, CNNs share these parameters across the entire layer. This parameter-sharing mechanism not only reduces the complexity of the network but also enhances its generalization capabilities. Furthermore, parameter sharing imparts translational invariance to the CNN, ensuring that the network's output remains consistent even when the input undergoes slight shifts or variations. This attribute is particularly advantageous in tasks such as image recognition, where CNNs have demonstrated exceptional performance. In recent years, parameter sharing has no longer been referred to solely as a compression technique; researchers and developers have started considering it a fundamental architecture within neural networks. Following the strategy of parameter sharing, the pruning technique is one of the most popular compression technologies. While Han et al. [38] introduced the method of pruning a model without losing the model accuracy. He found physiological evidence to support their pruning method. In mammalian physiology, it has been observed that during infancy, a large number of synaptic connections are formed. As the organism matures, the less frequently used synapses degrade and eventually disappear. Pruning steps were defined, and well-established research by him into three steps generally. First, the initial model is trained using standard methods, with the author positing that the magnitude of the weights indicates their importance. Next, weights below a certain threshold in the initial model are set to zero, effectively pruning the connections. Finally, the model is retrained to allow the remaining weights to compensate for any loss in accuracy caused by pruning. To achieve a satisfactory balance between compression ratio and accuracy, the pruning and retraining steps are repeated multiple times. To extend it, there are two main approaches: unstructured pruning and structured pruning. As illustrated in Figure 4.11, pruning nodes or neurons are generally categorized as structured pruning, whereas pruning individual weights/connections are classified as unstructured. Unstructured pruning operates at a finer granularity, allowing any proportion of redundant parameters to be removed without restriction. However, this can result in an irregular network structure post-pruning, potentially reducing the effectiveness of model acceleration. In other words, unstructured pruning selects parameters based on their importance rather than specific structural units. Conversely, structured pruning works at a coarser granularity, where the smallest pruning unit is a combination of parameters within a filter. By setting thresholds and evaluating the contribution of filters or feature maps, entire filters or certain channels below the threshold are removed, thus narrowing the network structure.

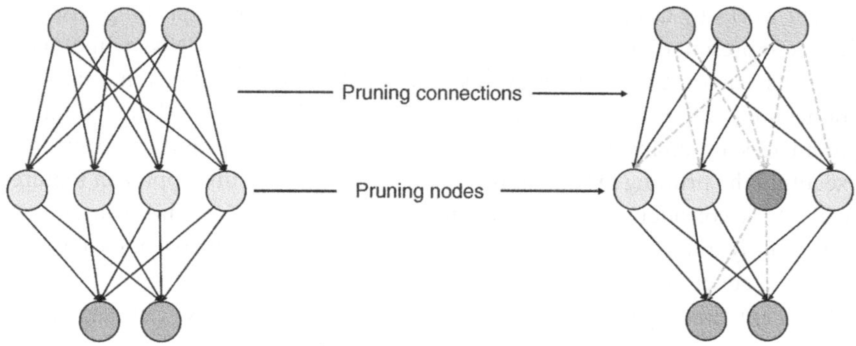

Figure 4.11 Overview of pruning.

This approach can achieve effective acceleration on existing software/hardware but may lead to a drop in model prediction accuracy. Therefore, fine-tuning the pruned model is necessary to compensate and restore its performance. Chen et al. [22] presented a HashedNets weight-sharing architecture that groups connection weights into hash buckets randomly by using a low-cost hash function, where all connections of each hash bucket have the same value. The values of the parameters are adjusted using the standard backpropagation method [110] during training. Han et al. [39] pruned redundant connections using a three-step method. First, the network learns which connections are important and then prunes the unimportant connections. Finally, they retrain the network to fine-tune the weights for the remaining connections.

In addition to the well-known unstructured and structured pruning methods, several other techniques have recently gained attention, although most are still categorized under structured pruning. One of them is filter pruning. Filter pruning focuses on removing the specific filters (or channels) of a neural network. The idea of filter pruning is that not all the channels equally contribute to the model's performance; some may be redundant or less important. This results in a more compact model with fewer parameters, which reduces both the memory footprint and computational requirement during inference. The most common approaches in filter pruning include the ranking filter based on their importance using metrics like L1/L2 norm [61], Taylor Expansion-based Pruning [84] or evaluating their removal on the loss function. Luo et al. [74] introduced a filter pruning based on the statistics information from the next layer named **Thinet**. Additionally, layer-wise pruning is a more aggressive approach than filter pruning. Instead of removing the individual filters, entire layers of the network are pruned. It eliminated not just the filters but also the associated computations and parameters for a whole layer. For instance, layers that contribute minimally to the final output or that have a high degree of redundancy might be pruned.

Recent works by Chen and Zhao [19] introduced a layer-wise pruning based on feature representation designed to reduce the complexity of CNNs while maintaining accuracy. Unlike the previous traditional pruning methods that focus on connection or filter-wise using weight information, they identified the redundant parameters by analyzing the features learned within convolution layers and executing the pruning process at the layer level. However, this approach requires meticulous consideration and evaluation because removing entire layers can significantly impact the network's architecture and potentially harm its ability to learn and generalize.

In recent times, pruning has not only been applied to CNNs but also, again, popularity to vision transformer (ViT) models. Hou and Kung [45] explored the multidimensional ViT compression approach that simultaneously targets redundancy reduction across the attention head, neuron, and sequence dimensions. They introduced a statistical dependence-based pruning criterion to identify and remove ineffective components across multiple dimensions. They then optimized the pruning strategy to maximize model accuracy within a computational budget, using an adapted Gaussian process search with expected improvement. Additionally, SP-ViT [119] incorporated a soft pruning method that reduced less informative tokens into a package token rather than removing them entirely, as identified by the selective module. It achieved significant computational results on vanilla transformers.

4.3.1.2 Quantization

Quantization with the rapid application of deep learning technology in various fields such as computer vision, natural language processing, and autonomous driving, a plethora of deep learning-based network models have emerged. However, these neural network models are large in parameters and complex in structure, making them suitable for inference on conventional GPUs but not for deployment on mobile and embedded devices. In real-world scenarios, these complex models often need to be deployed on low-cost embedded devices, creating a performance gap. Model quantization has emerged as a solution to effectively address this performance gap. Quantization is a model compression technique that converts floating-point storage (and computation) to integer storage (and computation). During training, complex and high-precision models are necessary to capture subtle gradient changes for optimization. However, high precision is not needed during inference as network parameters are fixed and no longer adjusted based on the loss function. Many parameters in the network are not critical or do not require high-precision representation. Moreover, experiments have shown that neural networks are robust to noise, and quantization can be considered as a form of noise. This means we can simplify the model before deployment by reducing the precision of representation. Most deep learning

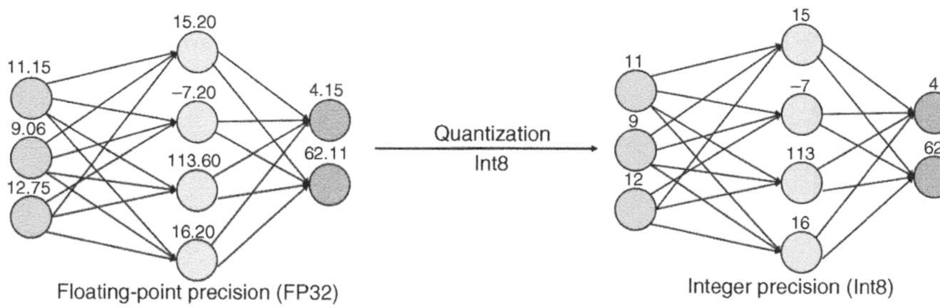

Figure 4.12 Overview of quantization.

training frameworks default to using 32-bit floating-point numbers for parameter representation and computation. The basic idea of model quantization is to replace the original floating-point precision with lower precision, such as 8-bit integers. Simply put, a weight that originally required a float 32 representation can be represented using Int8 after quantization. Figure 4.12 illustrates the overview of converting a neural network floating point precision to Int8 precision.

Current mainstream quantization methods are divided into linear quantization and nonlinear quantization. Linear quantization is the most commonly used method, particularly in the industry, where 8-bit quantization schemes are widely adopted. Linear quantization establishes a data mapping between high-precision floating-point values and low-precision fixed-point values. In nonlinear quantization, various "nonlinear" mapping functions are used, typically selected based on the characteristics of weight input distribution in different scenarios. A notable feature of nonlinear mapping is its ability to map weights of varying importance to different quantization ranges. For instance, if weight inputs are primarily distributed within a certain range, a nonlinear function can map these weights to a larger quantized range, enhancing the training process's sensitivity to the primary weight distribution. Another typical nonlinear quantization method involves using clustering techniques, such as k-means, during the initial model quantization phase. Weights are grouped into several clusters, and each cluster is quantized to the same fixed value to achieve the quantization effect. The foundational linear and nonlinear quantization methods can be performed using either quantization-aware training (QAT) or post-training quantization (PTQ) methods. Figures 4.13 and 4.14 illustrate the general workflow of the QAT and PTQ quantization. In general, QAT integrates into the neural network training process, allowing the model to adjust to the constraints of low-precision athematic while being trained. During QAT, the weights and activations are quantized during the forward pass, simulating the conditions under which the model will operate after deployment. The backward pass uses full precision (FP32) to ensure accurate gradient updates. The model is also fine-tuned to

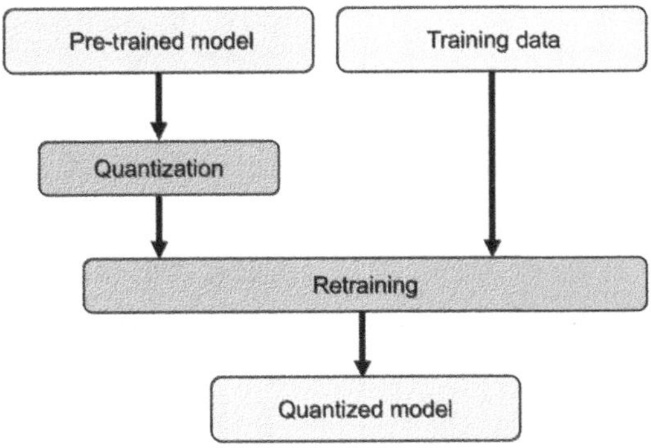

Figure 4.13 Overview of quantization-aware training (QAT).

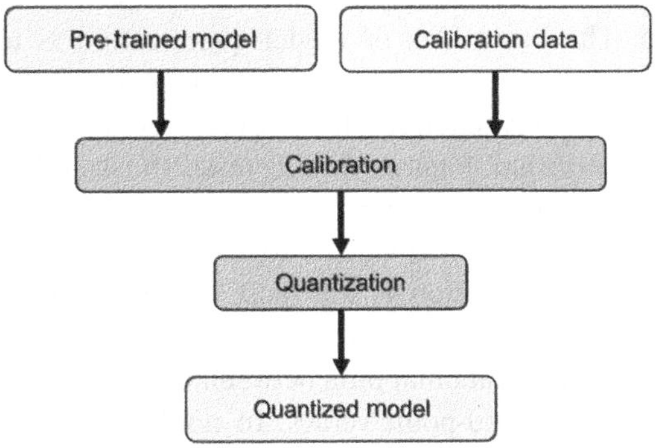

Figure 4.14 Overview of post-training quantization (PTQ).

minimize the accuracy loss, and quantized parameters are carefully calibrated to preserve performance. Contrarily, PTQ applies after the models have fully trained without retraining. This method often applies fine-tuning the quantized model using a calibration dataset (from training datasets) to mitigate potential accuracy loss. Both techniques can be utilized depending on the task one wants to perform or the types of hardware used. QAT usually requires more time due to a full training step, whereas PTQ is faster and more straightforward as it is applied after the model has been fully trained. PTQ is ideal when the original data is limited or unavailable. On the other hand, QAT learns to operate under quantization constraints, it can better handle the potential pitfalls of low-precision arithmetic, such as reduced dynamic range and quantization errors.

Courbariaux et al. [25] proposed a binary neural network to quantify the weights. More specifically, it restricts the value of the network weight by setting it to the value −1 or 1, and it simplifies the design of hardware that is dedicated to deep learning. Gong et al. [33] employed the *k*-means clustering algorithm to quantize the weights of fully connected layers, which could achieve up to 24 times the compression of the network with only a 1% loss of classification accuracy for the CNN network in the ImageNet challenge. Ding et al. [31]

introduced an accurate PTQ framework for vision transformers (APQ-ViT) that includes a unified block-wise calibration scheme to optimize quantization by addressing crucial errors on a block-by-block basis. Additionally, they introduced Matthew-Effect Preserving Quantization for the softmax function. Q-ViT [62] is one of the pioneers to introduce the QAT in the ViT. Q-ViT introduces an information rectification module (IRM) and a distribution guided distillation (DGD) scheme to reduce information distortion in the quantized self-attention map using low-bit quantization. Additionally, researchers have been more interested in applying mixed-precision quantization than single-precision in the full model. The idea behind this is to ensure higher precision in the sensitive layers (e.g., 16-bit or 8-bit) and lower precision in the less sensitive layers (e.g., 4-bit). For example, Tang et al. [102] proposed a novel method for mixed-precision quantization that utilized learnable scale factors as importance indicators to determine optimal bit-widths for each layer efficiently, significantly reducing search time and improving accuracy.

Lastly, quantization has not only been used in imaging; quantization is getting popular in compressing large language models (LLMs). The number of parameters of an LLM model is currently in billions and is expected to grow to trillions in the future. As a result, compressing the model is required to deploy LLMs in edge devices. Most of the LLMs currently use the PTQ technique as QAT can not scale up easily [65] in LLMs. Q-BERT [98] is the first work to apply quantization to the Bidirectional Encoder Representations from Transformers (BERT) model. They introduced a group-wise quantization and used a Hessian-based mix-precision technique to compress the model. One of the recent works by Xiao et al. [111] proposed a PTQ named Smoothquant for LLMs enabling 8-bit weight and 8-bit activations where they smooth the activation outliers, particularly focusing on mitigating the impact of outliers in activations and weights with a mathematically equivalent transformation. Extended to the Smoothquant work from the same MIT Han lab proposed AWQ [65] which minimizes the quantization error by preserving 1% of the salient weights. They use the activation distribution to identify the salient weight channels. This quantization technique is deployed on different edge devices, including Jetson Orin and Rasberry Pi4, and has also been experimented with on TinyChat.

4.3.1.3 Low-Rank Approximation

Low-rank approximation refers to reconstructing the dense matrix to estimate the representative parameters. If we consider the weight matrix of the original network as a full-rank matrix, we can use various low-rank approximation methods to decompose a large matrix multiplication into a series of multiplications between smaller matrices. This reduces the overall computation and accelerates the model's execution. First, let's understand what low rank and matrix rank mean. The rank of a matrix measures the linear independence of its rows and columns. A matrix is a full rank if all its rows and columns are linearly independent. The rank is determined by the number of nonzero rows or columns. In essence, the rank quantifies the matrix's inherent correlation. Consider an orchestra rehearsing a complex symphony. If each musician focuses on their own sheet music and their performances are well-coordinated, the entire piece will be harmonious, akin to a full-rank matrix where each part is linearly independent, with no redundancy. However, if some musicians ignore the conductor or disrupt the rhythm—for instance, if a violinist plays a different melody and others follow—then the orchestra's performance becomes chaotic. At this point, the orchestra resembles a low-rank matrix because parts of the performance are linearly dependent, losing their independence. This example illustrates that when all parts are independent and coordinated, the system (or matrix) has a high rank, is orderly, and problems are easily solved. Conversely, when parts influence each other and lose independence, the system's (or matrix's) rank decreases, leading to disorder and making problems harder to solve. In mathematics, the rank of a matrix is defined as the maximum number of linearly independent vectors within it, which can be understood as the degree of order. Since the rank measures correlation, the correlation within a matrix represents its structural information. If the rows of a matrix are highly correlated, the matrix can be projected into a lower-dimensional linear subspace, meaning it can be represented by a few vectors and is, therefore, low-rank. In summary, if a matrix represents structural information, such as images or user-item recommendation tables, it generally has a certain degree of correlation between its rows, making it typically low-rank. Regarding the low-rank estimation compression technique, some recent work was conducted. For example, Denton et al. [30] use singular value decomposition to reconstruct the weight of all connected layers. They triple the speedups of convolutional layers on both CPU and GPU, and the loss of precision is controlled within 1%. Denil et al. [28] employ low-rank approximation to compress the weights of different layers and reduce the number of dynamic parameters. Sainath et al. [96] uses a low-rank matrix factorization on the final weight layer of a DNN for acoustic modeling.

4.3.1.4 Knowledge Distillation

Knowledge distillation is also called teacher-student training. The idea of knowledge distillation is to adopt a teacher–student strategy and use a pretrained network to train a compact network for the same task [97]. It was first proposed by Caruana and coworkers [14]. They used a compressed network of trained network models to mark some unlabeled simulation data and reproduced the output of the original larger network. Figure 4.15 provides a visual summary of the

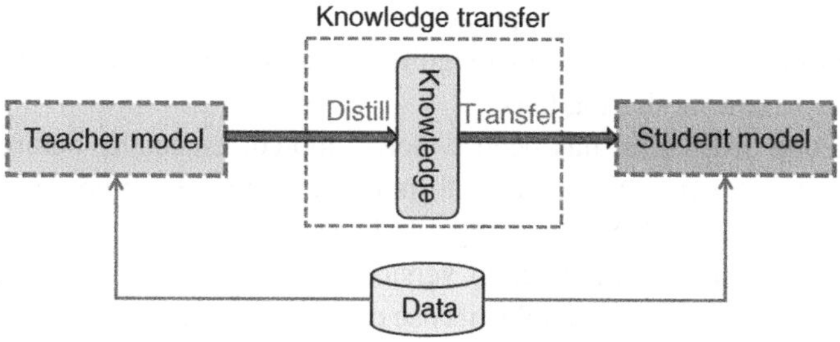

Figure 4.15 Overview of knowledge distillation.

knowledge distillation process within a neural network. It shows how knowledge is transferred from a Teacher Model to a Student Model. Knowledge distillation is primarily a technique used to transfer knowledge from one neural network to another, which can be either homogeneous or heterogeneous. The process begins by training a teacher network, which is then kept fixed during the distillation process. The output from this pretrained teacher network, along with the actual labels of the data, is used to train a student network. This two-step training process enables the student model to effectively learn the rich knowledge from the teacher model while also being refined using the ground truth labels. In constructing the loss function for knowledge distillation, one component is the Distill Loss, which is calculated as the cross-entropy between the soft targets from the teacher network and the softmax outputs of the student network. The other component is Student Loss, which is the cross-entropy between the ground truth labels and the softmax outputs of the student network. Knowledge distillation can be employed to compress a large network into a smaller one while retaining performance close to that of the larger network. Additionally, it can consolidate the knowledge learned by multiple networks into a single network, thereby achieving performance levels comparable to an ensemble of models. The work in [12] trained a parametric student model to estimate a Monte Carlo teacher model. Luo et al. [73] use the neurons in the hidden layer to generate more compact models and preserve as much of the label information as possible. Based on the idea of function-preserving transformations, the work in [20] instantaneously transfers the knowledge from a teacher network to each new, deeper, or wider network. However, using a shared hyperparameter between student and teacher requires a precise match in range and variance, which limits the performance of the student models due to the teacher's inherent logit patterns, which are already enough for effective learning. To overcome the limitations of shared hyperparameter in distillation, Sun et al. [101] introduced a hyperparameter based on the weighted standard deviation of logits and applying a Z-score standardization before softmax and Kullback–Leibler divergence. This approach allows the student to focus on key logit relationships rather than matching magnitudes, addressing issues where the traditional hyperparameter sharing fails.

Although knowledge distillation has been widely adopted, especially before deploying models, it has shown exceptional performance as a model compression technique for classification problems. Recently, LLMs have also garnered significant attention. For example, the lightweight versions of BERT, made possible through knowledge distillation, have enabled its use on edge devices. However, as larger models with more parameters emerge, there is a pressing need to explore new knowledge distillation techniques and develop innovative algorithms to effectively leverage these advanced models.

4.3.1.5 Combined Compression Techniques

Every compression technique has pros and cons or is sometimes unsuitable for the specific tasks to be performed in the hardware. As a result, researchers mixed different compression techniques and achieved better outcomes than the baseline results. Han et al. [38] coalesced pruning with quantization for model reduction and Huffman coding to reduce the storage requirement. Aghli and Ribeiro [1] combined weight pruning and knowledge distillation to compress the CNN models and solved the dimension dependencies of complex models like ResNets. They initially apply weight pruning selectively to specific layers within the CNN model to maintain the integrity of the network's structure. Following this, they implement knowledge distillation along with a customized loss function to further compress the layers that were not pruned, enhancing overall model efficiency. On the other hand, ViT also unified multiple compression techniques to achieve better accuracy with lower latency in edge devices. Yu et al. [113] addressed the high computational demands of ViTs by proposing a unified compression framework that combines layer-wise pruning, layer skipping, and knowledge distillation. Unlike existing approaches focusing on only one or two aspects of compression, this framework integrates all three techniques into an end-to-end, budget-constrained optimization process. The method jointly learns model weights, pruning ratios, and skip configurations under a distillation loss and is solved using the primal-dual algorithm. The experiment on ViT in ImageNet datasets can shrink to 50% of the original flops.

Knowledge distillation is mostly common when combining compression techniques because it provides a form of regularization that makes the student model more robust to the structural changes introduced by pruning or quantization. However, Qualcomm AI research [59] introduced Bayesian Bits for combining mixed-precision quantization and pruning. They use learnable stochastic gates to control bit widths, promoting low-bit solutions. As compression techniques continue to evolve, integrating methods like pruning, quantization, and knowledge distillation into unified frameworks represents a significant leap forward. These unified compression techniques not only push the boundaries of model efficiency but also ensure that deep learning models remain practical for real-world applications on edge devices.

4.3.2 Hardware-Software Codesign for Edge Optimization

Hardware-software codesign is a collaborative approach to system design in which software and hardware components are developed simultaneously. It maximizes edge devices' performance, efficiency, and scalability, balancing computational demand with available hardware resources. Figure 4.16 illustrates a hardware-software codesign's general workflow and components for efficiently deploying AI models into edge devices.

Tuli et al. [105] proposed a codesign framework named CODEBench for CNNs and their corresponding hardware accelerators. They addressed the limitation of search space and suboptimal exploration techniques. CODEBench consists of CNNBench (optimize CNN using a Bayesian second-order gradient search technique) and AccelBench (cycle-accurate simulations of hardware accelerator). CODEBench achieved higher top-1 accuracy, lower latency, and lower energy consumption on both ImageNet and Cifar datasets compared to the state-of-the-art pair on FPGA and Nvidla. Hardware-software codesign is explored and thriving in the CNN-based model and has also been successfully proposed for transformer-based models. For example, Zhou et al. [118] proposed an architecture named TransPIM that exemplifies software-hardware codesign by integrating processing-in-memory (PIM) techniques with transformer models. On the software level, it adopts a token-based dataflow to avoid inter-layer data flows; correspondingly, TransPIM incorporates lightweight modifications to the conventional high bandwidth memory architecture on the hardware level.

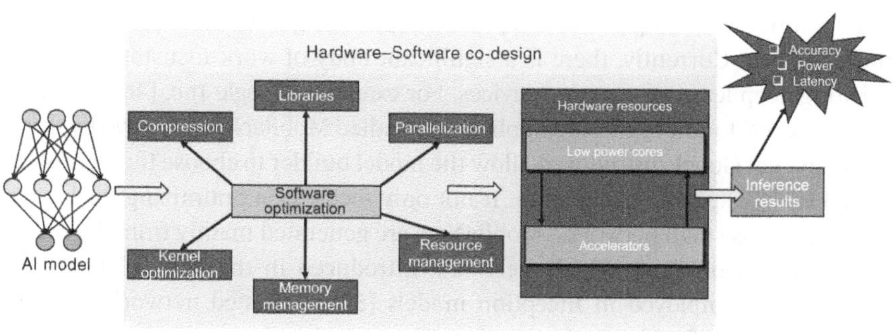

Figure 4.16 Overview of hardware-software codesign.

The overall results achieved 2.0× more throughput than existing ASIC-based accelerators.

Moreover, Hardware-software codesign on FPGA needs to map onto the FPGA's reconfigurable logic fabric, where custom PEs, such as systolic arrays or specialized arithmetic units, are instantiated to accelerate the different tasks such as object detection, image classification [2, 10, 108]. Li et al. [64] evaluated each module of the BEVDet (camera-based) and PointPillars (LiDAR-based) on FPGA and GPU to give insights about the necessity of the hardware-software codesign in the multimodal models. One of the recent works by Anupreetham et al. [10] proposed an end-to-end pipelined FPGA-based hardware-software codesign object detection system with 8× improvement in throughput. Wang et al. [108] explored the YOLOv2 model for CPU+FPGA platforms introducing a sparse convolution algorithm and FPGA accelerator architecture based on asynchronously executed parallel convolution cores. Although hardware-software on FPGA is evolving, it must be explored for real-time applications and resource-demanded models like transformers.

Additionally, Hardware-software codesign can be explored for multimodal multitask learning in autonomous systems. For instance, Hao and Chen [42] have comprehensive studies about the challenges, opportunities, and possible solutions in the future in the autonomous systems field.

4.3.3 Applying Deep Learning Models on Resource-Constrained Edges

First, let's explore what resource-constrained edge computing devices are. These devices have limitations in computational power, memory, storage, energy consumption, and network bandwidth. Typically, they are deployed in edge computing environments to process and analyze data generated near the source in real-time, rather than transmitting it to a remote data center for processing. As computational capabilities advance to support deep learning, the new task for edge computing is to perform deep learning training and inference directly on these devices. Currently, there is a significant body of work focusing on implementing deep learning on edge devices. For example, Google Inc. [46] presented efficient CNN for mobile vision applications, called MobileNets. The two hyperparameters that Google introduced allow the model builder to choose the right-sized model for the specific application. It not only focuses on optimizing for latency but also builds small networks. MobileNets are generated mainly from depthwise separable convolutions, which were first introduced in the work of [100] and subsequently employed in Inception models [50]. Flattened networks [53] are designed for fast feedforward execution. They consist of a consecutive sequence of one-dimensional filters that span every direction of three-dimensional space to achieve comparable performance as conventional convolutional networks [107]. Another small network is the Xception network [24]; Chollet et al. propose the dubbed Xception architecture inspired by Inception V3, where Inception modules have been replaced with depthwise separable convolutions. It shows that the architecture slightly outperforms Inception V3 on the ImageNet data set. Subsequently, Iandola et al. [47] developed Squeezenet, a small CNN architecture. It achieves AlexNet-level [56] accuracy with 50 times fewer parameters on the ImageNet data set (510 times smaller than AlexNet). In 2017, Microsoft Research India proposed Bonsai [57] and ProtoNN [37]. Then, they developed EMI-RNN [29] and FastGRNN [58] in 2018. Bonsai [57] refers to a tree-based algorithm used to efficiently predict devices in the IoT. More specifically, it is designed for supervised learning tasks such as regression, ranking, and multiclass classification. ProtoNN [37] is inspired by k-Nearest Neighbor (KNN) and could be deployed on the edges with limited storage and computational power (e.g., an Arduino UNO with 2 kB RAM) to achieve excellent prediction performance. EMI-RNN [29] requires 72 times less computation than standard Long Short-Term Memory Networks (LSTM) [44] and improves accuracy by 1%. Apple also developed efficient hybrid models named MobileViT [78] combining the strengths of both CNNs and ViTs to develop a lightweight and low latency network for mobile vision tasks. The main idea of MobileVit is to design transformers as convolutions in a way that the resultant MobileViT block has convolution-like properties while simultaneously allowing for global processing. MobileViT can successfully deploy on different mobile devices without extra effort (tested on iPhone 12). With fewer parameters, it performed 6.2% more accurately than MobileNetv3 and achieved 74.8% top-1 accuracy on the ImageNet dataset. Although MobileViT achieved high accuracy on mobile devices, it can not achieve low latency. Thereupon, apple proposed MobileViTv2 [79] and introduced a separable self-attention method with linear complexity to solve the bottleneck of multiheaded self-attention in transformers from MobileViT. This MobileViTv2 is the state-of-the-art for several mobile vision tasks, including object classification and detection.

However, due to its unique architecture, additional work such as hardware design, resource allocation, and scheduling is always required when deep learning models need to deploy on FPGA. Hardware acceleration is the most critical design in the workflow to facilitate the AI deployment of FPGA. Several works have been published that propose a hardware acceleration technique for FPGA. However, most have experimented with image classification and are still in the primary stage for real-world scenarios. For example, Auto-ViT-Acc [63] is one of the first works on ViT, which strategically utilized mixed precision across transformer blocks to match the computational demands and resource limitations of FPGAs. By allocating different bit-widths to PEs, the framework supported parallel processing of the ViT data flow and achieved 0.47%–1.36% higher

Top-1 accuracy under the same bit-width. Additionally, ViA [109] proposed a framework to overcome the significant processing power, path dependences, and memory bandwidth caused by ViT. ViA minimizes arisen path dependence from the model's shortcut mechanism, thereby optimizing the computational flow and securing efficient utilization of FPGA resources by employing a half-layer mapping strategy coupled with thorough throughput analysis. Furthermore, the architecture features two distinct reuse processing engines incorporating internal streams, diverging from traditional FPGA designs. The results indicated that it outperformed conventional computing platforms like NVIDIA's Tesla V100 in terms of energy efficiency, achieving approximately 5.2× better performance. Both works mainly focused on image classification tasks. In conclusion, edge computing has evolved significantly, enabling deep learning models to run efficiently on edge devices.

4.4 Edge Intelligent System Design and Optimization

Edge computing holds significant potential for expanding or even enhancing analytics capabilities that were previously limited to cloud environments [18]. In the meanwhile, given that intelligence is essential for quickly analyzing large data volumes and extracting insights, there is a growing demand to implement intelligence at the edge. Executing intelligent tasks near the data, rather than sending it to a remote server, enhances task efficiency and lowers the risk of data interception or leakage [72].

In this section, we focus on designing and optimizing EI systems, examining how to efficiently run AI models at the edge while addressing key factors such as algorithm performance, cost, privacy, reliability, and overall efficiency. Specifically, we delve into the implementation of EI, discussing key aspects such as (1) how to train models at the edge and (2) how to perform model inference at the edge.

4.4.1 Training on Edge

Model training plays a crucial role in setting the parameters of machine learning frameworks (such as neural networks) based on input data [66, 67]. Due to the limited computational capabilities of edge devices, this process has traditionally been performed off-device, often by sending data to a central server to free up computational resources for model inference. As a result, training machine learning models directly at edge nodes or servers is still relatively uncommon [70].

However, the concept of EI aims to leverage the data generated or collected by edge devices and train models locally rather than transmitting the data to a central server. This approach effectively addresses privacy and network concerns, providing a more secure and robust model training process that supports the development of practical AI services.

First, we focus on one of the most critical aspects of AI at the edge: how to train models at the network edge. We introduce the basic architecture of edge training, discuss optimization techniques, and explore federated learning, the most widely used method for this purpose.

4.4.1.1 Architecture for Model Training on Edge

- **Centralized Architecture:** Centralized architecture [71, 88], commonly referred to as client-server architecture, involves a system where a group of client edges requests and obtains services from a centralized server or cloud. In this model, the centralized server or cloud awaits service requests from the client edges and responds via a standardized interface. The client edges do not need to be aware of the specific details or configurations of the centralized server or cloud. This computing approach is particularly efficient when the client edges and the centralized server or cloud handle distinct, routine services.
- **Decentralized Architecture:** Decentralized architecture [71], also known as peer-to-peer (P2P) architecture, offers an alternative approach for communication and collaboration between edges and clouds. In this model, two edges can communicate and interact directly without involving a third party. Computing tasks are distributed among the edges, allowing them to both contribute and consume resources within the edge network, eliminating the need for a centralized server.

4.4.1.2 Optimization for Training

Once model training at the edge is adopted, it is essential to address the optimization challenges of the training process. The goal of optimization is to account for factors such as data distribution, computational power, and network capacity while ensuring that distributed edge deployment is feasible. Since solo training resembles the centralized architecture, our focus is primarily on collaborative training approaches. In this context, training optimization refers to improving the process to meet specific requirements, such as time efficiency, energy consumption, accuracy, and privacy protection. The main objective here is to accelerate the training process on resource-constrained edge devices, which can be approached in three key areas.

- **Processing efficiency:** It has been observed that the complexity of the training model significantly impacts time efficiency, especially when the device lacks sufficient computing resources [75]. To expedite the process, one approach is to reduce training time by using transfer learning, where learned features are transferred and cached locally for further training, thereby speeding up the overall process. Additionally, edge devices can collaborate and learn from each other, further enhancing training efficiency.

- **Communication efficiency:** To achieve communication efficiency, the focus is on reducing both the frequency and cost of communications. In other words, minimizing how often communication occurs and reducing the size of each communication exchange are key strategies for lowering communication costs. For instance, the authors in [68, 69] introduced collaborative training technologies on the edges for the AI-based prediction model and multitarget multiobject tracking. Besides the frequency of training updates, the size of these updates also impacts bandwidth usage. Gradient compression techniques, such as gradient quantization and gradient sparsification [15], can be employed to reduce the size of updates, thereby enhancing communication efficiency.

4.4.1.3 Collaborative Training

With the increasing computational and memory capabilities of edge devices, it raises the question of whether relying on the cloud for data processing is always necessary and whether innovative approaches can be implemented on the edge to address big data challenges. In response to these considerations, Lu et al. [68] propose a collaborative learning framework at the edge, called CLONE, which primarily demonstrates its effectiveness in reducing latency and preserving privacy (Figure 4.17).

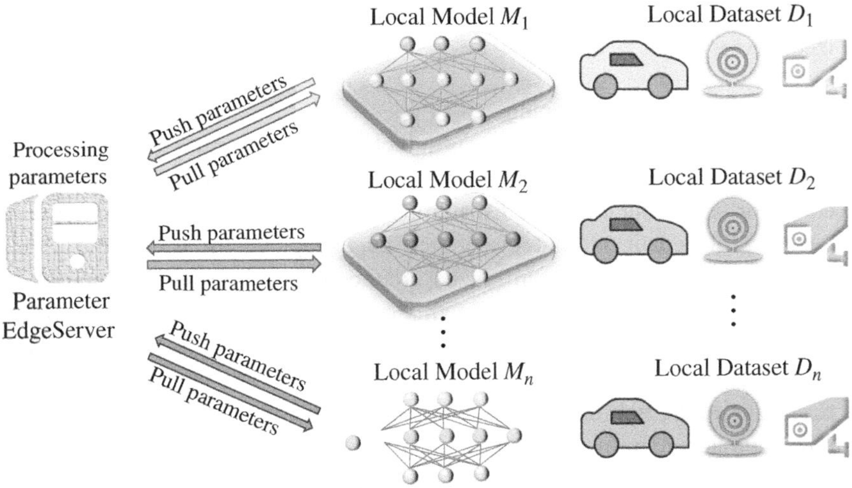

Figure 4.17 The framework of CLONE. The CLONE framework operates by allowing each edge node to locally train or run a neural network model using its own private data, while simultaneously sending its parameters to a central Parameter EdgeServer during the training or inference phase. The Parameter EdgeServer then performs necessary operations, such as aggregating the uploaded parameters, before transmitting the updated parameters back to the edge nodes. Source: Adapted from Lu et al. [68].

4.4.2 Model Inference on Edge

Edge inference is a key aspect of EI. As modern neural networks grow larger, deeper, and more complex, they demand increasingly substantial computing resources. This makes it challenging to run high-performance models directly on edge devices, such as mobile phones, IoT terminals, and embedded systems, which have limited computational power. Nevertheless, edge inference, as an essential part of EI, must be executed at the edge, where its overall performance (e.g., execution time, accuracy, and energy efficiency) can be significantly constrained by the device's capabilities. Here, we explore various frameworks and approaches aimed at bridging the gap between task requirements and device limitations.

4.4.2.1 Model Design

Recent studies have concentrated on developing lightweight neural network models that can be efficiently executed on edge devices with fewer hardware requirements. Based on the model design strategies, the existing literature can be grouped into two categories: architecture search and human-designed architecture. The former involves machines autonomously determining the optimal architecture, while the latter relies on human expertise to craft the architecture.

- **Human-designed architectures:** While architecture search has shown significant potential for model design, it continues to face challenges related to hardware requirements. As a result, researchers are increasingly focusing on human-designed strategies. For instance, they have developed lightweight deep neural networks, such as MobileNets [55], specifically for mobile and embedded devices by utilizing depth-wise separable convolutions. Another approach to reducing computational costs is the use of group convolutions, which has been employed to create foundational architectures like Xception [24].
- **Automotic architecture search:** Human-designed architectures are often time-consuming and require significant expertise. As a more efficient alternative, using AI to search for existing architectures and identify the optimal one for edge environments has gained traction. Automated search architectures like NASNet [120] and AmoebaNet [93] have demonstrated competitive, and sometimes superior, performance in tasks such as classification and recognition. However, despite the promising results of architecture search in model design, its popularity is still limited by the substantial hardware requirements it demands.

4.4.2.2 Efficient AI

Despite the widespread application and high performance of DNNs, their computational complexity remains a significant limitation, particularly for resource-constrained edge devices. High power consumption and latency can hinder system performance or even lead to crashes, as most edge devices are not built for compute-intensive tasks. Several approaches have been developed to address this issue. One approach is the design of specialized chips for deep learning, which accelerates tasks using dedicated hardware. Another solution is software-based, which involves evaluating whether all computations within the model are necessary. If not, the model can be simplified, reducing both the computational load and storage requirements.

Like other machine learning methods, DNNs consist of two phases: training and inference. During training, the model learns its parameters based on the training dataset, while in inference, the model uses test data to produce final results. Overparameterization refers to a scenario where numerous parameters are required during training to capture model fluctuations, but fewer are needed during inference. This allows the model to be simplified after training before being deployed at the edge for inference. Simplifying the model offers several advantages, including: (1) Reduced computation, leading to lower power consumption and shorter computation time. (2) A smaller memory footprint, allowing deployment on lower-end devices. More compact packages for application updates and releases. (3) This software-based technique is called model compression. It has a low implementation cost and is complementary to hardware acceleration, with the two methods potentially benefiting from each other. Model compression methods can be classified into four main categories: network pruning, quantization, knowledge distillation, and low-rank factorization.

4.4.2.3 Optimization Tool

Recently, researchers have turned their attention to TensorRT as a means to accelerate model inference while maintaining performance on resource-constrained edge devices (Figure 4.18). Sumaiya et al. [52] performed a comparative analysis of four different workflows using popular object detection models on TensorRT for Full Precision (FP32), Half Precision (FP16), and Integer Precision (INT8). Their findings highlight the inference performance and accuracy associated with each workflow. This chapter provides a comprehensive guide for selecting the most suitable workflow based on specific needs for inference performance and accuracy, offering valuable insights for advancements in edge devices (e.g., software-defined vehicles) and other real-time systems.

4.4.2.4 Collaborative Inference

Recently, researchers have also proposed collaborative inference frameworks for diverse edge computing applications. For example, as shown in Figure 4.19, Lu et al. [69] proposed a collaborative inference framework for multitarget multicamera tracking.

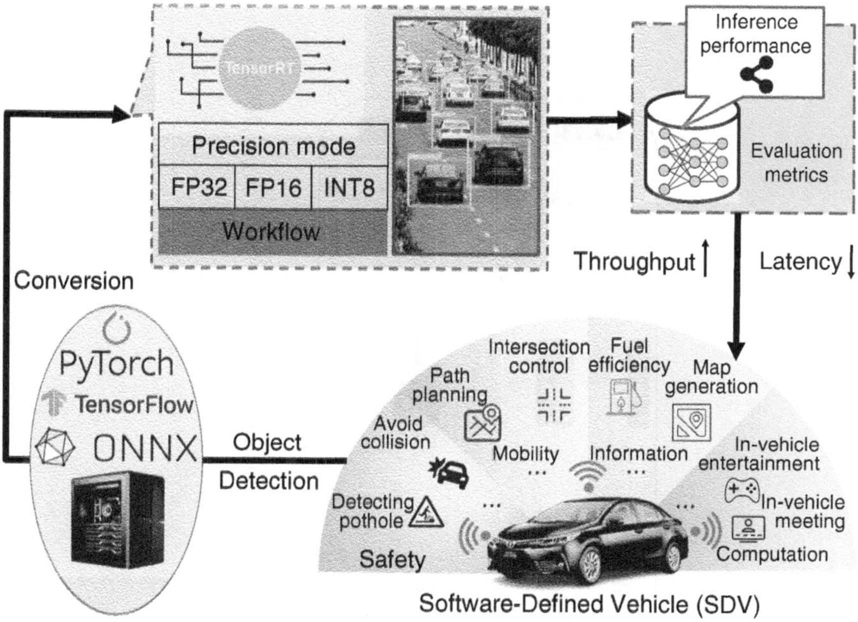

Figure 4.18 An overview of the TensorRT-enabled framework, which integrates popular object detection models from frameworks such as PyTorch, TensorFlow, and ONNX, along with NVIDIA GPUs, into TensorRT precision modes to optimize AI inference performance. Source: Jafarpourmarzouni et al. [52]/IEEE.

To be concrete, as shown in Figure 4.19, multiple video streams from different cameras (labeled Camera 1, Camera 2, and so on) are captured. This is the initial input where each camera provides its own set of video data. The video streams are passed through a target detection algorithm, such as YOLOv3 in this case, which identifies and outlines multiple targets (e.g., people) within the scene. Detected targets are represented by bounding boxes on the images. After target detection, features are extracted using wide residual networks to create an appearance descriptor (AD), which captures how the target looks (appearance features). Kalman filtering is applied to estimate and predict the target's motion over time, generating a motion descriptor (MD), which focuses on the movement of the target.

The detected and predicted appearance and MDs are then used to associate the detected targets with their identities across multiple frames and cameras. This data association is done using two main distance metrics: (1) Cosine distance: it is used to match the ADs across frames. (2) Mahalanobis distance: it is Used to match

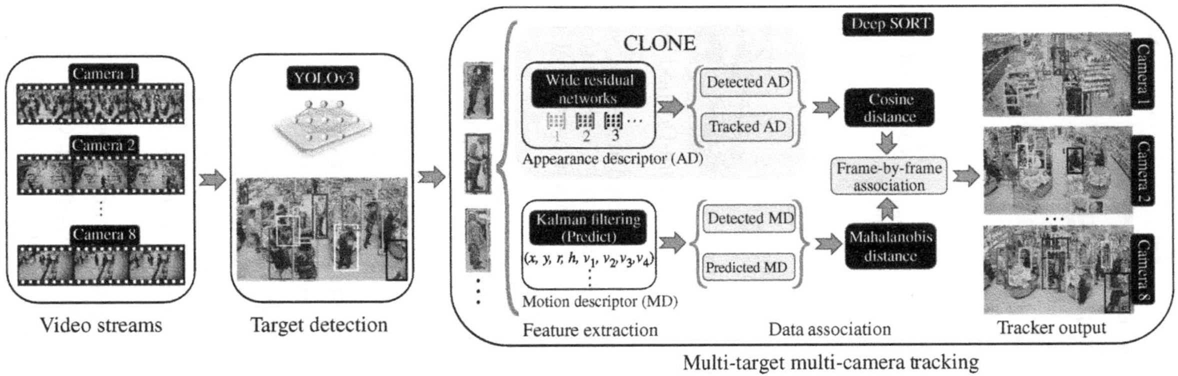

Figure 4.19 A depiction of the collaborative inference pipeline for the multitarget multicamera tracking. Source: Lu et al. [69]/IEEE.

the MDs based on their predicted movement trajectories. Finally, the targets are tracked across multiple cameras, with the corresponding bounding boxes highlighted on the camera feeds. This process allows the system to consistently track multiple individuals as they move across different camera views.

4.5 Summary and Practice

4.5.1 Summary

The rapid evolution of AI and edge computing has given rise to EI, a groundbreaking solution to the challenges posed by the enormous data volumes generated in our interconnected world. By shifting data processing to the network's edge, this innovative approach significantly reduces bandwidth consumption and associated costs while maintaining high-quality services, offering a compelling alternative to traditional cloud-based processing. In this chapter, we've defined EI as the ability of edge devices to execute AI algorithms locally. This capability extends to processing a wide array of data types, including video streams, natural language, time-series information, and unstructured sensor data, without the need for cloud uploads. To enable EI, a diverse ecosystem of technologies has emerged. On the hardware front, specialized components such as ASICs, FPGAs, and GPUs have been developed to accelerate AI tasks while minimizing power consumption. Complementing these advancements, software frameworks like TensorFlow Lite, CoreML, XNNPACK, and QNNPACK have been optimized for edge environments, facilitating efficient AI inference on devices with limited resources. We've also explored the critical role of collaboration between cloud and edge systems, as well as edge-to-edge interactions. These collaborative approaches are essential for distributed deep learning and real-time processing, enabling models to be trained in the cloud and deployed efficiently at the edge, with the added benefit of local fine-tuning based on edge-generated data. Various compression techniques play a pivotal role in adapting complex AI models for edge deployment. Methods such as parameter sharing, pruning, quantization, and low-rank approximation effectively reduce model size and computational demands, making it feasible to run sophisticated AI models on resource-constrained edge devices. Furthermore, we've highlighted knowledge distillation as a powerful technique for transferring insights from large, pretrained models to smaller, more efficient versions suitable for edge deployment. This process ensures that the streamlined models maintain high accuracy while optimizing for low latency, energy efficiency, and minimal memory usage. Moreover, we highlighted the unified compression techniques in both CNN and ViT models to discuss the future scope of the model compression techniques. Subsequently, hardware-software codesign is also crucial in optimizing performance and balancing computational demands with hardware capabilities in devices like FPGAs. Recent works have demonstrated significant improvements in throughput and efficiency through this codesign approach, such as in YOLOv2 implementations on CPU+FPGA platforms. Accelerating. Moreover, from early models like MobileNets and SqueezeNet to more recent hybrid models like MobileViT and MobileViTv2, the emphasis has shifted toward directly balancing accuracy, latency, and efficiency to meet the demands of real-time applications without any additional huddle. On the other hand, as the need for deploying those complex models on specialized hardware like FPGAs grows, hardware-accelerating techniques such as Auto-ViT-Acc and ViA are emerging to tackle the unique challenges posed by ViTs. Both software and hardware design architecture developments are needed to expedite the AI deployment on edge devices. In essence, EI represents the convergence of AI and edge computing, providing a robust framework for deploying advanced AI capabilities directly at the network edge. By leveraging cutting-edge hardware, optimized software, collaborative processing, and model compression techniques, EI promises to revolutionize AI applications across diverse domains, from smart home systems to autonomous vehicles, enhancing both performance and efficiency.

4.5.2 Practice Questions

1) How does edge computing hardware differ from traditional data center hardware?
2) Discuss the advantages and challenges of implementing machine learning at the edge.
3) What are the key considerations in choosing an edge application development framework?
4) How does integrating hardware accelerators impact the design of machine learning models for edge deployment?
5) Discuss the role of software-hardware codesign in optimizing resource-constrained edge computing environments.

4.5.3 Course Projects

1) Develop a simple edge computing application using a containerization platform.
2) Prune a classical neural network model to reduce both model size and latency. Understand the basic concept of pruning, implement and apply a few pruning approaches, get a basic understanding of performance improvement (such as

speedup) from pruning, and understand the differences and tradeoffs between these pruning approaches.

3) Quantize a classical neural network model to reduce both model size and latency. Understand the basic concept of quantization, implement and apply a few quantization approaches, get a basic understanding of performance improvement (such as speedup) from quantization, and understand the differences and tradeoffs between these quantization approaches.
4) Use knowledge distillation to compress a classical neural network model to reduce both model size and latency. Understand the basic concept of knowledge distillation and get a basic understanding of performance improvement (such as speedup) from knowledge distillation.
5) Using model compression techniques, optimizing large language models (LLMs) on edge devices (e.g., your laptop). A good example can be found at Github: https://github.com/mit-han-lab/tinychat-tutorial?tab=readme-ov-file.

Chapter 4 Suggested Papers

1 Jude Haris et al. "SECDA: Efficient hardware/software co-design of FPGA-based DNN accelerators for edge inference". In: *2021 IEEE 33rd International Symposium on Computer Architecture and High Performance Computing (SBAC-PAD)*. IEEE. 2021, pp. 33–43.

2 Jakub Konečný et al. "Federated learning: Strategies for improving communication efficiency". In: *arXiv preprint arXiv:1610.05492* (2016).

3 Xingzhou Zhang et al. "OpenEI: An open framework for edge intelligence". In: *2019 IEEE 39th International Conference on Distributed Computing Systems (ICDCS)*. IEEE. 2019, pp. 1840–1851.

4 Zhi Zhou et al. "Edge intelligence: Paving the last mile of artificial intelligence with edge computing". In: *Proceedings of the IEEE* 107. 8 (2019), pp. 1738–1762.

References

1 Nima Aghli and Eraldo Ribeiro. "Combining weight pruning and knowledge distillation for CNN compression". In: *Proceedings of the IEEE/CVF Conference on Computer Vision and Pattern Recognition*. 2021, pp. 3191–3198.

2 Afzal Ahmad, Muhammad Adeel Pasha, and Ghulam Jilani Raza. "Accelerating tiny YOLOv3 using FPGA-based hardware/software co-design". In: *2020 IEEE International Symposium on Circuits and Systems (ISCAS)*. IEEE. 2020, pp. 1–5.

3 Amazon Web Services. *AWS Snowball.* https://aws.amazon.com/cn/snowball/. Accessed: 2024-07-09. 2024.

4 Amazon Web Services. *AWS Snowcone.* https://aws.amazon.com/cn/snowcone/. Accessed: 2024-07-09. 2024.

5 Amazon Web Services. *AWS IoT Greengrass.* https://aws.amazon.com/greengrass/. Accessed: 2024-07-10. 2024.

6 Amazon Web Services, Inc. *Amazon SageMaker Neo: Train Once, Run Anywhere.* https://aws.amazon.com/sagemaker/neo/. Accessed: 2024-08-21. 2024.

7 Amazon Web Services, Inc. *Amazon SageMaker Edge.* https://aws.amazon.com/sagemaker/edge/. Accessed: 2024-08-20. 2024.

8 Amazon Web Services, Inc. *AWS RoboMaker.* https://aws.amazon.com/cn/robomaker/. Accessed: 2024-01-20. 2024.

9 AMD Xilinx. *Vitis AI.* https://www.xilinx.com/products/design-tools/vitis/vitis-ai.html. 2024. (Visited on 08/20/2024).

10 Anupreetham Anupreetham et al. "High throughput FPGA-based object detection via algorithm-hardware co-design". In: *ACM Transactions on Reconfigurable Technology and Systems* 17. 1 (2024), pp. 1–20.

11 Apple Inc. *Core ML Documentation.* https://developer.apple.com/documentation/coreml. Accessed: 2024-05-20. 2024.

12 Anoop Korattikara Balan et al. "Bayesian dark knowledge". In: *Advances in Neural Information Processing Systems* 28 (2015).

13 Saman Biookaghazadeh, Ming Zhao, and Fengbo Ren. "Are *FPGAs* suitable for edge computing?" In: *USENIX Workshop on Hot Topics in Edge Computing (HotEdge 18).* 2018.

14 Cristian Buciluă, Rich Caruana, and Alexandru Niculescu-Mizil. "Model compression". In: *Proceedings of the 12th ACM SIGKDD International Conference on Knowledge Discovery and Data Mining.* 2006, pp. 535–541.

15 Yi Cai et al. "Long live time: Improving lifetime for training-in-memory engines by structured gradient sparsification". In: *Proceedings of the 55th Annual Design Automation Conference.* 2018, pp. 1–6.

16 Canonical Ltd. *LXD - The system container manager.* https://canonical.com/lxd. Accessed: 2024-08-14. 2024.

17 Jie Cao et al. "EdgeOSH: A home operating system for internet of everything". In: *2017 IEEE 37th International Conference on Distributed Computing Systems (ICDCS).* IEEE. 2017, pp. 1756–1764.

18 Junzhou Chen and Sidi Lu. "An advanced driving agent with the multimodal large language model for autonomous vehicles". In: *2024 IEEE International Conference on Mobility, Operations, Services and Technologies (MOST).* IEEE. 2024, pp. 1–11.

19 Shi Chen and Qi Zhao. "Shallowing deep networks: Layer-wise pruning based on feature representations". In: *IEEE Transactions on Pattern Analysis and Machine Intelligence* 41. 12 (2018), pp. 3048–3056.

20 Tianqi Chen, Ian Goodfellow, and Jonathon Shlens. "Net2net: Accelerating learning via knowledge transfer". In: *arXiv preprint arXiv:1511.05641* (2015).

21 Tianqi Chen et al. "MxNet: A flexible and efficient machine learning library for heterogeneous distributed systems". In: *arXiv preprint arXiv:1512.01274* (2015).

22 Wenlin Chen et al. "Compressing neural networks with the hashing trick". In: *International Conference on Machine Learning*. PMLR. 2015, pp. 2285–2294.

23 Yu Cheng et al. "A survey of model compression and acceleration for deep neural networks". In: *arXiv preprint arXiv:1710.09282* (2017).

24 Francois Chollet. "Xception: Deep learning with depthwise separable convolutions". In: *Proceedings of the IEEE Conference on Computer Vision and Pattern Recognition*. 2017, pp. 1251–1258.

25 Matthieu Courbariaux, Yoshua Bengio, and Jean-Pierre David. "Binary Connect: Training deep neural networks with binary weights during propagations". In: *Advances in Neural Information Processing Systems* 28 (2015).

26 Mike Davies et al. "Loihi: A neuromorphic manycore processor with on-chip learning". In: *IEEE Micro* 38. 1 (2018), pp. 82–99.

27 DDS. *DDS Foundation*. https://www.dds-foundation.org/. Accessed: 2024-01-10. 2024.

28 Misha Denil et al. "Predicting parameters in deep learning". In: *Advances in Neural Information Processing Systems* 26 (2013).

29 Don Dennis et al. "Multiple instance learning for efficient sequential data classification on resource-constrained devices". In: *Advances in Neural Information Processing Systems* 31 (2018).

30 Emily L Denton et al. "Exploiting linear structure within convolutional networks for efficient evaluation". In: *Advances in Neural Information Processing Systems* 27 (2014).

31 Yifu Ding et al. "Towards accurate post-training quantization for vision transformer". In: *Proceedings of the 30th ACM International Conference on Multimedia*. 2022, pp. 5380–5388.

32 Zidong Du et al. "ShiDianNao: Shifting vision processing closer to the sensor". In: *Proceedings of the 42nd Annual International Symposium on Computer Architecture*. 2015, pp. 92–104.

33 Yunchao Gong et al. "Compressing deep convolutional networks using vector quantization". In: *arXiv preprint arXiv:1412.6115* (2014).

34 Google. *XNNPACK*. https://github.com/google/XNNPACK. GitHub repository. 2024. (Visited on 08/20/2024).

35 Google Cloud. *Edge TPU*. https://cloud.google.com/edge-tpu. Accessed: 2024-07-09. 2024.

36 Google Cloud. *Google Kubernetes Engine*. https://cloud.google.com/kubernetes-%3Cp%3Eengine?%3Cp%3Ehl%3Cp%3E=en. Accessed: 2024-08-14. 2024.

37 Chirag Gupta et al. "ProtoNN: Compressed and accurate KNN for resource-scarce devices". In: *International Conference on Machine Learning*. PMLR. 2017, pp. 1331–1340.

38 Song Han, Huizi Mao, and William J Dally. "Deep compression: Compressing deep neural networks with pruning, trained quantization and Huffman coding". In: *arXiv preprint arXiv:1510.00149* (2015).

39 Song Han et al. "Learning both weights and connections for efficient neural network". In: *Advances in Neural Information Processing Systems* 28 (2015).

40 Song Han et al. "EIE: Efficient inference engine on compressed deep neural network". In: *ACM SIGARCH Computer Architecture News* 44. 3 (2016), pp. 243–254.

41 Song Han et al. "ESE: Efficient speech recognition engine with sparse LSTM on FPGA". In: *Proceedings of the 2017 ACM/SIGDA International Symposium on Field-Programmable Gate Arrays*. 2017, pp. 75–84.

42 Cong Hao and Deming Chen. "Software/hardware co-design for multi-modal multi-task learning in autonomous systems". In: *2021 IEEE 3rd International Conference on Artificial Intelligence Circuits and Systems (AICAS)*. IEEE. 2021, pp. 1–5.

43 Jason Hill et al. "System architecture directions for networked sensors". In: *ACM SIGPLAN Notices* 35. 11 (2000), pp. 93–104.

44 Sepp Hochreiter and Jürgen Schmidhuber. "Long short-term memory". In: *Neural Computation* 9. 8 (1997), pp. 1735–1780.

45 Zejiang Hou and Sun-Yuan Kung. "Multi-dimensional model compression of vision transformer". In: *2022 IEEE International Conference on Multimedia and Expo (ICME)*. 2022, pp. 01–06. DOI: 10.1109/ICME52920. 2022.9859786.

46 Andrew G Howard et al. "MobileNets: Efficient convolutional neural networks for mobile vision applications". In: *arXiv preprint arXiv:1704.04861* (2017).

47 Forrest N Iandola et al. "SqueezeNet: AlexNet-level accuracy with 50x fewer parameters and < 0.5 MB model size". In: *arXiv preprint arXiv:1602.07360* (2016).

48 Intel. *OpenVINO Toolkit Overview*. https://www.intel.com/content/www/us/en/developer/tools/openvino-toolkit/overview.html. Accessed: 2024-07-10. 2024.

49 International Electrotechnical Commission. *Edge Intelligence*. https://www.iec.ch/basecamp/edge-intelligence. Accessed: 2024-05-27. 2024.

50 Sergey Ioffe and Christian Szegedy. "Batch normalization: Accelerating deep network training by reducing internal covariate shift". In: *International Conference on Machine Learning*. PMLR. 2015, pp. 448–456.

51 iRobot Corporation. *iRobot: Robot Vacuums and Mops*. https://www.irobot.com/. Accessed: 2024-01-20. 2024.

52 Sumaiya et al. "Enhancing real-time inference performance for time-critical software-defined vehicles". In: *2024 IEEE International Conference on Mobility, Operations, Services and Technologies (MOST)*. IEEE. 2024, pp. 101–113.

53 Jonghoon Jin, Aysegul Dundar, and Eugenio Culurciello. "Flattened convolutional neural networks for feedforward acceleration". In: *arXiv preprint arXiv:1412.5474* (2014).

54 Norman P Jouppi et al. "In-datacenter performance analysis of a tensor processing unit". In: *Proceedings of the 44th Annual International Symposium on Computer Architecture*. 2017, pp. 1–12.

55 Whui Kim, Woo-Sung Jung, and Hyun Kyun Choi. "Lightweight driver monitoring system based on multi-task mobilenets". In: *Sensors* 19. 14 (2019), p. 3200.

56 Alex Krizhevsky, Ilya Sutskever, and Geoffrey E Hinton. "ImageNet classification with deep convolutional neural networks". In: *Advances in Neural Information Processing Systems* 25 (2012).

57 Ashish Kumar, Saurabh Goyal, and Manik Varma. "Resource-efficient machine learning in 2 KB RAM for the Internet of Things". In: *International Conference on Machine Learning*. PMLR. 2017, pp. 1935–1944.

58 Aditya Kusupati et al. "FastGRNN: A fast, accurate, stable and tiny kilobyte sized gated recurrent neural network". In: *Advances in Neural Information Processing Systems* 31 (2018).

59 Andrey Kuzmin et al. "Pruning vs quantization: Which is better?" In: *Advances in Neural Information Processing Systems* 36 (2024).

60 Philip Levis et al. "TinyOS: An operating system for sensor networks". In: *Ambient Intelligence* (2005), pp. 115–148.

61 Hao Li et al. "Pruning filters for efficient convnets". In: *arXiv preprint arXiv:1608.08710* (2016).

62 Yanjing Li et al. "Q-ViT: Accurate and fully quantized low-bit vision transformer". In: *Advances in Neural Information Processing Systems* 35 (2022), pp. 34451–34463.

63 Zhengang Li et al. "Auto-ViT-Acc: An FPGA-aware automatic acceleration framework for vision transformer with mixed-scheme quantization". In: *2022 32nd International Conference on Field-Programmable Logic and Applications (FPL)*. IEEE. 2022, pp. 109–116.

64 Yunge Li, Shaibal Saha, and Lanyu Xu. "The architectural implications of multi-modal detection models for autonomous driving systems". In: *2024 IEEE International Conference on Mobility, Operations, Services and Technologies (MOST)*. 2024, pp. 218–228. DOI: 10.1109/MOST60774.2024.00030.

65 Ji Lin et al. "AWQ: Activation-aware weight quantization for on-device LLM compression and acceleration". In: *Proceedings of Machine Learning and Systems* 6 (2024), pp. 87–100.

66 Sidi Lu and Weisong Shi. "The emergence of vehicle computing". In: *IEEE Internet Computing* 25. 3 (2021), pp. 18–22.

67 Sidi Lu and Weisong Shi. "Vehicle computing: Vision and challenges". In: *Journal of Information and Intelligence* 1. 1 (2023), pp. 23–35.

68 Sidi Lu, Yongtao Yao, and Weisong Shi. "Collaborative learning on the edges: A case study on connected vehicles". In: *2nd USENIX Workshop on Hot Topics in Edge Computing (HotEdge 19)*. 2019.

69 Sidi Lu, Yongtao Yao, and Weisong Shi. "CLONE: Collaborative learning on the edges". In: *IEEE Internet of Things Journal* 8. 13 (2020), pp. 10222–10236.

70 Sidi Lu et al. "SafeCampus: Multimodal-based campus-wide pandemic forecasting". In: *IEEE Internet Computing* 26. 1 (2021), pp. 60–67.

71 Sidi Lu et al. "A comparison of end-to-end architectures for connected vehicles". In: *2022 5th International Conference on Connected and Autonomous Driving (MetroCAD)*. IEEE. 2022, pp. 72–80.

72 Sidi Lu et al. "EdgeWare: Toward extensible and flexible middleware for connected vehicle services". In: *CCF Transactions on High Performance Computing* 4. 3 (2022), pp. 339–356.

73 Ping Luo et al. "Face model compression by distilling knowledge from neurons". In: *Proceedings of the AAAI Conference on Artificial Intelligence*. Vol. 30. 1. 2016.

74 Jian-Hao Luo, Jianxin Wu, and Weiyao Lin. "ThiNet: A filter level pruning method for deep neural network compression". In: *Proceedings of the IEEE International Conference on Computer Vision*. 2017, pp. 5058–5066.

75 Yichen Luo et al. "Impact of raindrops on camera-based detection in software-defined vehicles". In: *2024 IEEE International Conference on Mobility, Operations, Services and Technologies (MOST)*. IEEE. 2024, pp. 193–205.

76 Steven Macenski et al. "Robot operating system 2: Design, architecture, and uses in the wild". In: *Science Robotics* 7. 66 (2022), eabm6074.

77 Dukhan Marat, W Yiming, and L Hao. *QNNPACK: Open source library for optimized mobile deep learning*. 2018.

78 Sachin Mehta and Mohammad Rastegari. "MobileViT: Light-weight, general-purpose, and mobile-friendly vision transformer". In: *arXiv preprint arXiv:2110.02178* (2021).

79 Sachin Mehta and Mohammad Rastegari. "Separable self-attention for mobile vision transformers". In: *arXiv preprint arXiv:2206.02680* (2022).

80 Dirk Merkel. "Docker: Lightweight linux containers for consistent development and deployment". In: *Linux Journal* 2014. 239 (2014), p. 2.

81 Microsoft. *Azure IoT Edge*. https://azure.microsoft.com/en-us/products/iot-edge. Accessed: 2024-07-10. 2024.

82 Microsoft Azure. *Azure Sphere*. https://azure.microsoft.com/en-us/products/azure-sphere. Accessed: 2024-07-09. 2024.

83 Dharmendra S Modha. "Introducing a brain-inspired computer". In: Published online at http://www.research.ibm.com/articles/brain-chip.shtml (2017).

84 Pavlo Molchanov et al. "Importance estimation for neural network pruning". In: *Proceedings of the IEEE/CVF Conference on Computer Vision and Pattern Recognition*. 2019, pp. 11264–11272.

85 NVIDIA Corporation. *NVIDIA Jetson Orin*. https://www.nvidia.com/en-us/autonomous-machines/embedded-systems/jetson-orin/. Accessed: 2024-05-20. 2024.

86 NVIDIA Corporation. *TensorRT*. https://developer.nvidia.com/tensorrt. Accessed: 2024-05-20. 2024.

87 NVIDIA Corporation. *Jetson Xavier Series*. https://www.nvidia.com/en-us/autonomous-machines/embedded-systems/jetson-xavier-series/. Accessed: 2024-05-20. 2024.

88 Cristina Olaverri-Monreal. "Autonomous vehicles and smart mobility related technologies". In: *Infocommunications Journal* 8. 2 (2016), pp. 17–24.

89 ONNX Runtime Developers. *ONNX Runtime*. https://onnxruntime.ai/. 2021.

90 George Plastiras et al. "Edge intelligence: Challenges and opportunities of near-sensor machine learning applications". In: *2018 IEEE 29th International Conference on Application-Specific Systems, Architectures and Processors (ASAP)*. IEEE. 2018, pp. 1–7.

91 PX4. *PX4 Autopilot*. https://px4.io/. Accessed: 2024-01-10. 2024.

92 Morgan Quigley et al. "ROS: An open-source Robot Operating System". In: *ICRA Workshop on Open Source Software*. Vol. 3. 3.2. Kobe, Japan. 2009, p. 5.

93 Esteban Real et al. "Regularized evolution for image classifier architecture search". In: *Proceedings of the AAAI Conference on Artificial Intelligence*. Vol. 33. 01. 2019, pp. 4780–4789.

94 Red Hat. *What is rkt?* https://www.redhat.com/en/topics/containers/what-is-rkt. Accessed: 2024-08-14. 2024.

95 ROBOTIS. *ROBOTIS Official Website*. http://www.robotis.us. Accessed: 2024-01-20. 2024.

96 Tara N Sainath et al. "Low-rank matrix factorization for deep neural network training with high-dimensional output targets". In: *2013 IEEE International Conference on Acoustics, Speech and Signal Processing*. IEEE. 2013, pp. 6655–6659.

97 Bharat Bhusan Sau and Vineeth N Balasubramanian. "Deep model compression: Distilling knowledge from noisy teachers". In: *arXiv preprint arXiv:1610.09650* (2016).

98 Sheng Shen et al. "Q-BERT: Hessian based ultra low precision quantization of BERT". In: *Proceedings of the AAAI Conference on Artificial Intelligence* 34. 05 (2020), pp. 8815–8821.

99 Weisong Shi et al. "Edge computing: Vision and challenges". In: *IEEE Internet of Things Journal* 3. 5 (2016), pp. 637–646.

100 Laurent Sifre and Stéphane Mallat. "Rigid-motion scattering for texture classification". In: *arXiv preprint arXiv:1403.1687* (2014).

101 Shangquan Sun et al. "Logit standardization in knowledge distillation". In: *Proceedings of the IEEE/CVF Conference on Computer Vision and Pattern Recognition (CVPR)*. 2024, pp. 15731–15740.

102 Chen Tang et al. "Mixed-precision neural network quantization via learned layer-wise importance". In: *European Conference on Computer Vision*. Springer. 2022, pp. 259–275.

103 Surat Teerapittayanon, Bradley McDanel, and Hsiang-Tsung Kung. "Distributed deep neural networks over the cloud, the edge and end devices". In: *2017 IEEE 37th International Conference on Distributed Computing Systems (ICDCS)*. IEEE. 2017, pp. 328–339.

104 TensorFlow Authors. *TensorFlow Lite*. https://www.tensorflow.org/lite. Accessed: 2024-05-21. 2024.

105 Shikhar Tuli et al. "CODEBench: A neural architecture and hardware accelerator co-design framework". In: *ACM Transactions on Embedded Computing Systems* 22. 3 (2023), pp. 1–30.

106 Yaman Umuroglu et al. "FINN: A framework for fast, scalable binarized neural network inference". In: *Proceedings of the 2017 ACM/SIGDA International Symposium on Field-Programmable Gate Arrays*. 2017, pp. 65–74.

107 Min Wang, Baoyuan Liu, and Hassan Foroosh. "Factorized convolutional neural networks". In: *Proceedings of the IEEE International Conference on Computer Vision Workshops*. 2017, pp. 545–553.

108 Zixiao Wang et al. "Sparse-YOLO: Hardware/software co-design of an FPGA accelerator for YOLOv2". In: *IEEE Access* 8 (2020), pp. 116569–116585.

109 Teng Wang et al. "ViA: A novel vision-transformer accelerator based on FPGA". In: *IEEE Transactions on Computer-Aided Design of Integrated Circuits and Systems* 41. 11 (2022), pp. 4088–4099.

110 Paul J Werbos. "Backpropagation through time: What it does and how to do it". In: *Proceedings of the IEEE* 78. 10 (1990), pp. 1550–1560.

111 Guangxuan Xiao et al. "SmoothQuant: Accurate and efficient post-training quantization for large language models". In: *International Conference on Machine Learning*. PMLR. 2023, pp. 38087–38099.

112 Xilinx Inc. *PYNQ (Python Productivity for Zynq)*. http://www.pynq.io. Accessed: 2024-08-21. 2024.

113 Shixing Yu et al. "Unified visual transformer compression". In: *arXiv preprint arXiv:2203.08243* (2022).

114 Qingyang Zhang et al. "OpenVDAP: An open vehicular data analytics platform for CAVs". In: *2018 IEEE 38th International Conference on Distributed Computing Systems (ICDCS)*. IEEE. 2018, pp. 1310–1320.

115 Shaojun Zhang et al. "Enabling edge intelligence for activity recognition in smart homes". In: *2018 IEEE 15th International Conference on Mobile Ad Hoc and Sensor Systems (MASS)*. IEEE. 2018, pp. 228–236.

116 Xingzhou Zhang, Yifan Wang, and Weisong Shi. "pCAMP: Performance comparison of machine learning packages on the edges". In: *USENIX Workshop on Hot Topics in Edge Computing (HotEdge 18)*. 2018.

117 Xingzhou Zhang et al. "OpenEI: An open framework for edge intelligence". In: *2019 IEEE 39th International Conference on Distributed Computing Systems (ICDCS)*. IEEE. 2019, pp. 1840–1851.

118 Minxuan Zhou et al. "TransPIM: A memory-based acceleration via software-hardware co-design for transformer". In: *2022 IEEE International Symposium on High-Performance Computer Architecture (HPCA)*. IEEE. 2022, pp. 1071–1085.

119 Yuxuan Zhou et al. "SP-ViT: Learning 2D spatial priors for vision transformers". In: *arXiv preprint arXiv:2206.07662* (2022).

120 Barret Zoph et al. "Learning transferable architectures for scalable image recognition". In: *Proceedings of the IEEE Conference on Computer Vision and Pattern Recognition*. 2018, pp. 8697–8710.

5

Challenges and Solutions in Edge Computing*

While implementing the potential applications of edge computing, it is essential to address the challenging key technical difficulties inherent in these applications. To realize the vision of edge computing, researchers and developers in computer systems, networks, and application services need to engage in close collaboration and communication. This chapter summarizes several critical issues that urgently need to be addressed in edge computing research. It proposes some solutions and research directions worth further exploration based on existing research results.

5.1 Programmability and Data Management

The efficiency and flexibility of edge systems are highly related to programmability and data management. This section explores the challenges of making edge platforms programmable, focusing on the complexities of automatic program partitioning, naming conventions, and data abstraction. Solving these issues enables develop systems that are not only easier to program but also more adept at managing the vast amount of data generated at the edge.

5.1.1 Programmability

In the cloud computing model, users write applications and deploy them to the cloud. Cloud service providers maintain the cloud servers, and users typically know little or nothing about the operation of these programs. This transparency of infrastructure is a key advantage of application development in the cloud computing model. User programs are usually written and compiled for the target platform and run on cloud servers.

*This chapter is contributed by Sidi Lu.

Edge Computing: Systems and Applications, First Edition. Lanyu Xu and Weisong Shi.

Companion Website: www.wiley.com/go/xu/computing

However, in the edge computing model, part or all of the computing tasks are migrated from the cloud to edge nodes. Since edge nodes are mostly heterogeneous platforms and each node's runtime environment may differ, deploying user applications in the edge computing model presents significant challenges for programmers. Traditional programming methods such as MapReduce [10] and Spark [53] are unsuitable, necessitating research into new programming methods based on edge computing.

To achieve programmability in edge computing, Zhang et al. proposed a programming model based on hybrid cloud and edge computing, known as the Firework model [55]. In the era of the Internet of Everything, this model addresses the need for distributed sharing and processing of big data when data production and consumption are both migrated to edge devices. It also enables the functionality of computation flows in edge computing. Computation flow refers to the series of computations that can be performed on data along its transmission path by edge nodes, allowing data to be incrementally processed, reducing the amount of data transmitted.

As shown in Figure 5.1, the Firework model consists of two types of nodes: Firework Model Managers and Firework Model Nodes. It defines datasets and functions through a virtual shared data view, integrating geographically distributed data sources. Data stakeholders (Firework Model Nodes) provide a set of predefined function interfaces for end-users to access. When using the Firework system, users can focus more on business implementation, while the communication, function scheduling, and composition can be managed

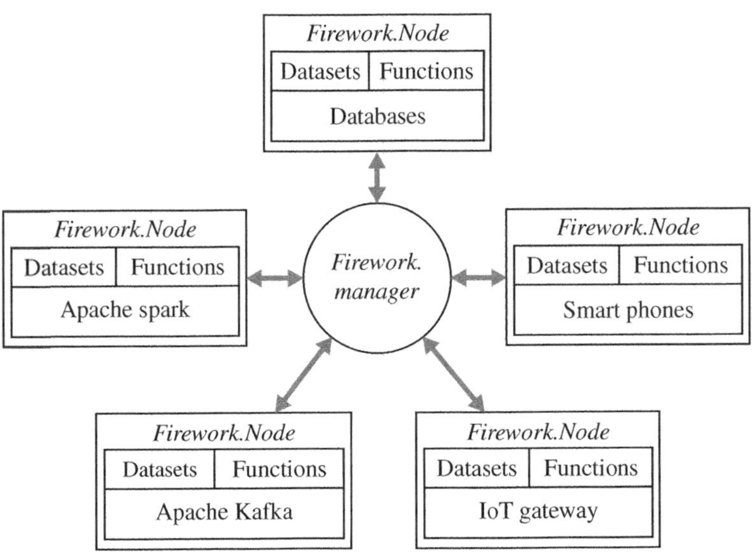

Figure 5.1 Edge computing paradigm.

using the programming interfaces provided by Firework. By deploying and configuring Firework Model Nodes and their mutual functions, the system achieves distributed sharing and processing of big data and supports the functionality of computation flows.

The Firework model extends the visualization boundary of data, proposing a new programming paradigm for distributed data processing in collaborative edge environments. Each participant in the Firework model can achieve data processing on local devices and the integration of cloud and edge computing resources. Additionally, it is important to note that the collaborative issues in the edge computing model (such as synchronization, data/state migration, etc.) are among the pressing problems in programmability that need to be addressed.

Satyanarayanan and coworker proposed the OpenStack++ model [18], which is primarily applied in the cloudlet architecture. It provides a programming model for application developers tailored for mobile environments. Amento et al. proposed the FocusStack model [1], which supports the deployment of diverse and complex applications on various potential IoT edge devices. Edge devices are constrained in terms of computing, power consumption, and connectivity, and they are highly mobile. FocusStack first identifies edge devices with sufficient resources, then deploys and runs applications on these devices. This model allows developers to focus solely on program design while the FocusStack model determines the appropriate edge devices and tracks their status. Sajjad et al. proposed the SpanEdge model [41], which unifies cloud central nodes and near-edge central nodes, reducing network latency caused by wide-area network connections. It provides a programming environment where developers can focus on developing stream processing applications, specifying which parts of the application need to run near the data source without worrying about the data source and its geographical distribution.

5.1.2 Automatic Program Partitioning

In the edge computing environment, as the computational capability of edge nodes improves, migrating applications from cloud centers to edge nodes becomes a significant challenge. Distributing originally standalone applications across different network edge nodes is crucial for the feasibility and efficiency of application design in edge computing systems. This process directly impacts the executability and efficiency of edge computing applications. Designing and implementing partitioning techniques for applications to ensure the reasonable allocation of application components among cloud-edge and edge-edge heterogeneous nodes is essential for achieving high performance and reliability in edge computing environments.

Program partitioning in edge computing environments needs to consider various state information, such as resources, energy consumption, and response latency of edge nodes, to decompose applications into multiple components while preserving the original application's semantics. These components are then placed onto different nodes. The existing partitioning methods mainly include static and dynamic partitioning: static program partitioning is completed during the compilation process, commonly seen in message passing interface (MPI) programming and heterogeneous multicore programming based on general-purpose graphics processing unit (GPGPU) computing cards; dynamic program partitioning is primarily performed during the runtime.

The edge computing environment shares certain similarities with distributed environments, allowing programs to be designed, implemented, and debugged at the central node to ensure they can run on edge nodes. However, the partitioned programs need to be distributed across various edge nodes. For homogeneous edge nodes, this is similar to program partitioning in a distributed system environment. Nevertheless, edge computing environments often have heterogeneous nodes, making traditional partitioning methods insufficient for edge computing needs. These traditional methods do not consider the characteristics of the edge computing environment, such as resource heterogeneity, varying data sources, and edge node mobility. Therefore, program partitioning in edge computing environments, in addition to static and dynamic partitioning, needs to address the specificities of cloud-edge and edge-edge partitioning.

The concept of the program dependence graph (PDG) [12] can be leveraged, retaining the original program component nodes while adding features like heterogeneous resource availability weights, location distance parameters, and communication costs between edge nodes in the edge computing environment. Furthermore, in addition to the original dependency relations (such as data dependency and control dependency), the PDG can incorporate dependencies specific to the edge computing environment, including resource availability dependency, location mobility dependency, and response time dependency. These dependencies can be utilized in the dependency analysis process, applicable not only in the static compilation phase and dynamic runtime phase but also in the cloud-edge and edge-edge program partitioning. This ensures that applications can be reasonably allocated to different edge nodes while guaranteeing the reliability and high performance of the execution of different components on the edge nodes.

5.1.3 Naming Conventions

A significant assumption in the edge computing model is the enormous number of edge devices. At the edge nodes, numerous applications operate, each with its own framework for service delivery. As with any computing system, the naming scheme in edge computing holds significant importance for programming, addressing, device identification, and data communication. Yet, an efficient and standardized naming mechanism tailored for the edge computing paradigm has yet to be established. Edge practitioners often must familiarize themselves with diverse communication and network protocols to interact with the heterogeneous devices within their systems. The naming scheme for edge computing must accommodate device mobility, highly dynamic network topologies, privacy and security concerns, and scalability to manage the immense volume of unreliable devices effectively.

Traditional naming mechanisms like domain name system (DNS) and uniform resource identifiers perform well in current networks. However, they often lack the flexibility needed to support dynamic edge networks, where many devices may be highly mobile and resource-constrained. Additionally, IP-based naming schemes can be too cumbersome for resource-constrained edge devices due to their complexity and overhead.

New naming mechanisms such as Named Data Networking (NDN) [54] and MobilityFirst [40] have been proposed for edge computing. NDN offers a hierarchically structured naming scheme suited for content-centric networks, enhancing scalability and human-friendly service management at the edge. However, integrating NDN with other protocols like Bluetooth or ZigBee requires additional proxies and raises security concerns regarding hardware information isolation.

MobilityFirst addresses mobility support by separating names from network addresses, which is beneficial for highly mobile edge environments. However, its requirement for a globally unique identifier (GUID) for naming may not be practical for fixed-edge environments like home networks. Moreover, GUIDs are not user-friendly for service management.

For smaller, fixed-edge environments such as homes, a solution could involve the edge operating system (edgeOS) assigning a network address to each device based on a unique human-friendly name. As shown in Figure 5.2, this approach includes information about location, role, and data description in the name (e.g., "kitchen.oven2.temperature3"). This human-friendly naming convention simplifies service management, device diagnosis, and component replacement. Users and service providers can easily understand notifications from the edgeOS, facilitating quick actions without the need for error codes or network reconfiguration. Furthermore, this naming approach enhances programmability for service providers while safeguarding hardware information, thus improving data privacy and security. It allows the mapping of human-friendly names to unique identifiers used for edgeOS management and network addresses (e.g., IP addresses or MAC addresses) for supporting various communication protocols like Bluetooth, ZigBee, or WiFi.

Addressing highly dynamic edge environments at a city-wide scale remains an ongoing challenge, requiring further investigation and community collaboration.

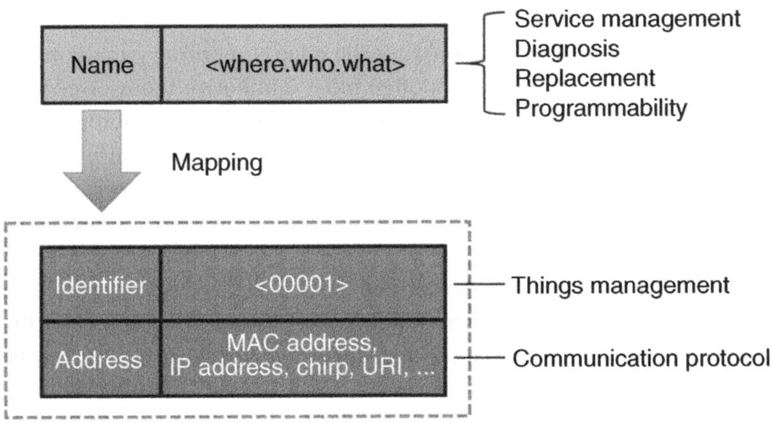

Figure 5.2 The naming mechanism for the edge operating system (edgeOS). Source: Shi et al. [43]/IEEE.

5.1.4 Data Abstraction

Various applications running on the edgeOS interact with the service management layer through wireless communication, relying on position indicators. Data abstraction has been extensively studied in wireless sensor networks and cloud computing paradigms. However, edge computing introduces new challenges due to the vast number of IoT devices generating data, such as in a smart home environment. Here, nearly all devices continuously report data to the edgeOS, scattered throughout the home. For instance, a thermometer might report temperature every minute, while a security camera records video sent to the gateway, often without immediate consumption before being replaced by newer footage.

Besides, applications in edge computing systems use data or provide services through service management layer APIs. Compared to cloud computing, data abstraction in edge computing is more challenging. In a smart home, intelligent devices, as data producers, send data to the edge computing system. However, there are fewer devices deployed around the home, and most network edge devices periodically send sensed data to a gateway. For instance, a temperature sensor sends temperature data to the gateway every minute, though it is used infrequently. Based on this, Shi et al. [43] proposed reducing human involvement in edge computing by having edge nodes process the data and interact with users proactively. In this scenario, the gateway layer preprocesses the data (e.g., denoising, event detection, and privacy protection) before sending it to the upper layers of the system as source data for application services. This process faces the following three challenges:

- **Diversity of data formats from different devices**: Due to considerations of data privacy and security, source data is transparent to the tasks running on the gateway. These tasks should extract the information needed for processing from an integrated data table. Shi et al. [43] proposed a table structure containing numbered, named, timestamped, and data fields, allowing edge device data to be stored in this table. However, this hides the details of the sensed data, affecting its usage.

- **Uncertainty in the degree of data abstraction**: If data abstraction filters out too much source data, some applications or services may fail due to insufficient information. Conversely, retaining a large amount of source data poses a challenge for system developers in terms of data storage and management. Additionally, the data sent by edge devices is often unreliable. Extracting useful information from unreliable sources remains a technical challenge.
- **Applicability of data abstraction**: Edge devices collect data and provide it for application use, completing specific services. These applications should have read and write permissions for devices to cater to user-specific needs. The data abstraction layer combines data representation and operations, providing a common interaction interface for devices connected to the edge computing system. Due to the heterogeneity of edge devices, data representation and operations vary, creating a barrier to general data abstraction.

Given the practical requirements of real-world applications, researchers aim to reduce human intervention in edge computing by enabling edge nodes to preprocess data and engage proactively with users. At the gateway level, preprocessing tasks encompass noise reduction, event detection, and privacy protection before transmitting processed data to upper layers for service delivery. However, several challenges accompany this strategy. Firstly, data from diverse devices arrives in varying formats (refer to Figure 5.3). To uphold privacy and security, gateway applications should access unified data tables without exposure to raw data specifics, using a standardized format (e.g., 0000, 12:34:56 PM, 01 January 2016,

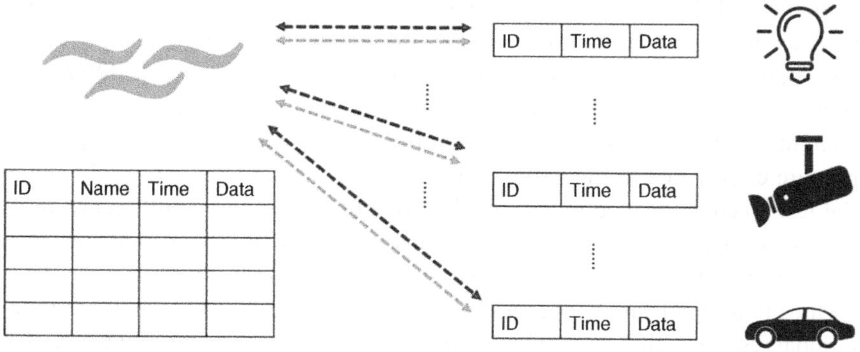

Figure 5.3 Data abstraction in edge computing scenarios. Source: Shi et al. [43]/IEEE.

kitchen.oven2.temperature3, 78). Nevertheless, abstracting data at this level risks reducing its practical utility.

Secondly, determining the appropriate level of data abstraction is complex. Filtering too much raw data may limit application learning capabilities, whereas storing vast amounts of raw data poses storage challenges. Moreover, data reliability concerns persist due to sensor inaccuracies, hazardous environments, and unreliable wireless connections, posing ongoing challenges for IoT developers seeking to extract useful insights from potentially unreliable data sources.

Another critical aspect of data abstraction involves enabling applications to perform operations on connected devices. Data abstraction layers serve as public interfaces for these operations, accommodating the heterogeneous nature of connected devices with diverse data representations and operational capabilities, thereby complicating universal data abstraction efforts.

5.2 Resource Allocation and Optimization

Efficient resource allocation and optimization are critical to deploying edge computing systems successfully. With the ever-increasing demand for low-latency processing and real-time data handling, scheduling strategies, data offloading, and load balancing play a pivotal role. This section delves into these optimization challenges, highlighting the need for intelligent resource management to ensure that edge computing systems meet performance expectations while remaining scalable and adaptable.

5.2.1 Scheduling Strategies

The scheduling strategies in edge computing aim to optimize resource utilization, reduce response time, minimize energy consumption, and enhance the overall performance of task processing in the edge computing environment. Compared to traditional distributed systems, the scheduling strategies of edge computing systems share some similarities, such as the distributed handling of computational tasks and resources across various nodes. However, there are notable differences, such as the heterogeneity of computational resources, which is more akin to cloud computing systems. Unlike cloud computing, edge computing scheduling strategies are closely tied to their specific computing environments, primarily due to the resource-constrained nature of edge computing systems. Additionally, these strategies must consider the overhead caused by the mobility of users, which is different from the scheduling strategies in cloud computing systems.

One of the significant challenges in edge computing is how to schedule computational resources effectively. The scheduling strategy in edge computing is related to resources, ensuring that the resources used by a specific application during its execution are efficiently managed. Given the heterogeneity of data, computation, storage, and network resources in edge computing task scheduling, it is necessary to design heterogeneous resource scheduling strategies tailored to different application instances. Moreover, the diversity of applications requires scheduling strategies that can support various types of applications, ensuring their normal operation. These strategies should maximize the utilization of limited computational resources to enhance the executability and efficiency of applications in the edge computing environment while minimizing resource usage. For edge computing resource providers or service providers, the scheduling strategy should also maximize resource benefits. Thus, real-time monitoring and tracking of application execution and resource changes are needed to achieve dynamic scheduling of applications and the resources required for their execution.

Existing research indicates that scheduling strategies in edge computing environments can be implemented using graph theory methods [3]. Specifically, each application is represented by a graph structure where each node represents a component of the application, and the edges between nodes represent communication between them. Physical resources can also be represented by a graph, with nodes representing computational resources (such as servers) and edges representing the relationships between them. This approach transforms the resource scheduling problem into a mapping relationship between resource nodes and applications. In terms of resource scheduling, parts of user applications can run on the central cloud or on resources at the network edge. In edge computing environments, resource availability, network conditions, and user locations are dynamically changing. Thus, parts of an application may need to migrate from one edge node/central cloud server to another. An optimal scheduling strategy for an application must consider network status and user mobility.

Professor Qun Li's research team [52] has focused on the issue of edge computing response latency, providing a solution for task scheduling between edge nodes. Their approach primarily addresses the state of task execution between edge nodes using three strategies: shortest transmission time first, shortest queue length first, and shortest scheduling delay first. These strategies are all related to delay time, leveraging one of the advantages of edge computing—reducing data transmission latency. Designing and implementing a scheduling strategy that effectively reduces task execution delay at edge nodes is one of the challenges encountered in the research on edge computing scheduling strategies.

5.2.2 Data Offloading and Load Balancing

An edge computing-based system is inherently a distributed system [37, 39], leveraging diverse data sources and distributed computational capabilities. When individual edge devices or servers reach their processing limits, task partitioning and data offloading becomes essential. Effective load balancing becomes more

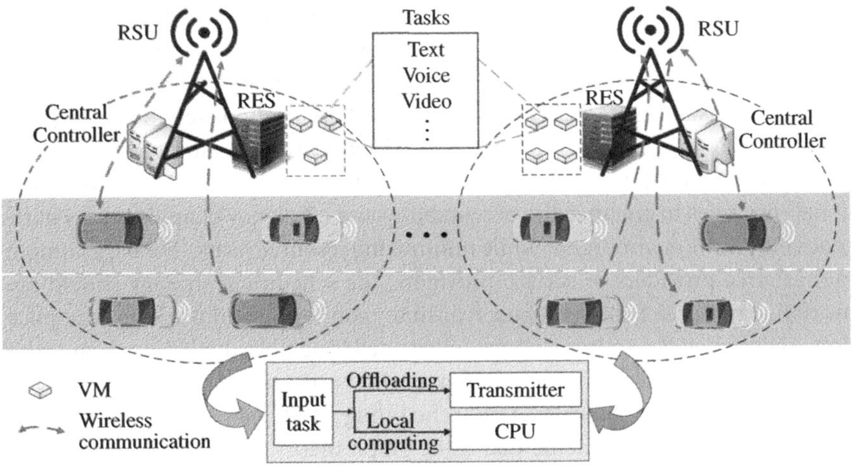

Figure 5.4 Offloading framework in edge computing scenarios for vehicle-edge collaboration. Source: Luo et al. [30]/IEEE.

critical as device performance decreases, especially in scenarios like vehicle-edge collaboration [30] (as shown in Figure 5.4). This involves partitioning tasks and data across multiple devices or servers, and aggregating results to achieve overall system load balancing. Load balancing and data offloading optimize computing and storage resources across layers, preventing single-resource overload, reducing latency, and enhancing the efficiency, reliability, availability, and scalability of the edge computing-based system redundantly. However, challenges persist in data offloading and load balancing in edge computing that require ongoing research and improvement.

5.2.2.1 Data Offloading

In conventional edge computing systems, data offloading typically focuses on large servers, with individual devices often overlooked. Conversely, in edge computing-based edge computing systems, every device capable of processing data necessitates consideration for data offloading, significantly increasing the complexity of this issue. Moreover, due to the constrained computing and storage capacities of edge devices, frequent data offloading is required to balance the overall system load. This results in high throughput in the edge network, intensive bandwidth usage, and potential delays in task execution. To address these challenges, tailored data offloading schemes must be devised based on specific requirements. Generally, depending on offloading needs and application contexts, two modes of data offloading can be employed: full offloading and partial offloading.

(i) **Full Data Offloading**: When considering full offloading, factors such as time delay and energy consumption play crucial roles. If devices have sufficient energy and there is a stringent delay requirement, the offloading scheme should prioritize minimizing delay. Optimal solutions involve leveraging task queue information, resource utilization of edge nodes and servers, and routing status to determine the most efficient full data offloading strategy [27]. Additionally, optimizing energy consumption through an optimization framework while meeting delay constraints is paramount [24].

(ii) **Partial Data Offloading**: Unlike full offloading which focuses on time delay and energy consumption, partial offloading involves dividing data from a task into blocks, with only selected blocks being offloaded [7]. Decisions on which blocks to offload depend on various parameters such as total data volume, device resource utilization, channel conditions, and energy consumption [33]. These parameters are carefully considered to jointly optimize the allocation of communication and computing resources. Despite advancements, current offloading strategies in edge devices require further adaptation and optimization to align with the specific characteristics of industrial environments.

5.2.2.2 Load Balancing

When designing a data offloading scheme, considerations typically revolve around minimizing both time delay and energy consumption. However, such schemes may inadvertently overload certain devices, necessitating effective load-balancing strategies [31]. The primary goal of load balancing is to ensure equitable distribution of workload across edge nodes and maintain stable communication links, thereby optimizing the utilization of computing and network resources. Given the scale and frequency of operations in edge systems, it becomes imperative to enhance existing load-balancing algorithms to align with the unique characteristics of these environments.

In edge computing-based systems, task and load data are collected from far-edge, mid-edge, and near-edge devices and servers, organized hierarchically, and integrated with artificial intelligence (AI) to develop load balancing mechanisms tailored to each layer. Furthermore, SDN (software-defined networking) can be leveraged to orchestrate load balancing across the entire edge network, simplifying scheduling and routing complexities.

5.2.3 Optimization Metrics

In the edge computing model, different layers have varying computational capabilities, making load distribution a critical issue. It is necessary to determine which layer should handle specific loads or how loads should be distributed across each layer. Various allocation strategies are typically employed, such as evenly distributing loads across all layers according to the number of layers, or assigning the maximum load to a single layer. In extreme cases, all tasks might be allocated to either the edge or the cloud. Several optimization metrics should be considered when selecting the optimal load distribution strategy, including latency, bandwidth, energy consumption, and cost [43].

5.2.3.1 Latency

Latency is one of the most important metrics for evaluating performance [8, 11], especially in interactive applications. Cloud servers offer high computational power, completing complex tasks like image and speech recognition in a short time. However, latency is influenced not only by computation time but also by transmission time. Long delays in wide-area networks can significantly impact real-time or interaction-intensive applications [6]. To reduce latency, loads should be executed at the nearest layer with computational capability. For instance, in a smart city, users can preprocess photos on their local devices before sending information about a missing person to the cloud, avoiding the need to upload all photos to the cloud. The closest physical layer might not always be the optimal choice. It is necessary to avoid unnecessary waiting times by considering resource usage to find a reasonable optimization layer. For example, if a user is playing a game that occupies a lot of the phone's computational resources, uploading photos through the nearest gateway would be more efficient.

As depicted in Figure 5.5, Xu et al. [48] proposed ChatCache, a scalable edge system that incorporates a hierarchical cache design to serve both single and multiple users. On most evaluated platforms, ChatCache significantly reduces user-perceived latency by over 91.7% for voice requests and more than 81.6% for text requests.

5.2.3.2 Bandwidth

From a latency perspective, high bandwidth can reduce transmission delays, especially for large data transfers [9]. For short-distance transmissions, future research can explore high-bandwidth wireless access technologies to send data to the edge. If the edge can handle tasks, it significantly improves latency and saves transmission bandwidth between the edge and the cloud. In smart homes, gateways can handle most data through WiFi or other high-speed transmission methods. Short

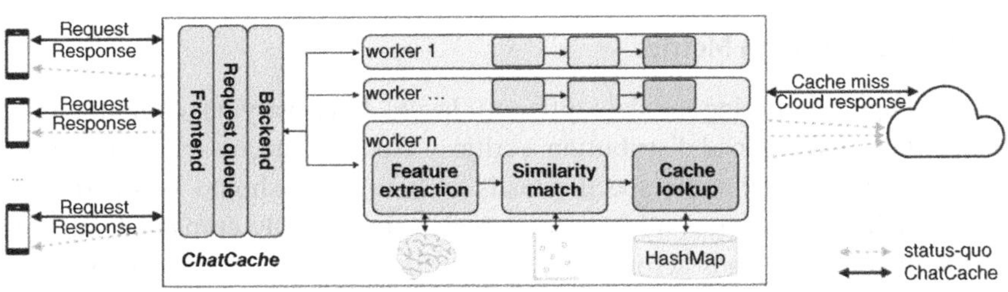

Figure 5.5 The design of ChatCache. Source: Xu et al. [48]/IEEE.

transmission paths also improve data transfer reliability. If edge devices cannot meet computational requirements, they can preprocess source data to reduce the upload volume significantly. In a smart-city scenario, local preprocessing of photos before uploading can save bandwidth and reduce user data transmission. Globally, the saved bandwidth can be used for other edge user data uploads and downloads. When using high bandwidth in edge computing, it is essential to evaluate the appropriate speed configuration for the edge. Additionally, to avoid competition and latency, the computational capacity and bandwidth at each layer must be considered for load distribution.

5.2.3.3 Energy Consumption

The battery is the most constrained resource for edge devices. For terminal devices, offloading tasks to the edge layer can save energy [42]. For a specific load, determining whether migrating the entire or part of the load to the edge layer is energy-efficient requires balancing computation and transmission energy consumption. Generally, it is necessary to determine whether the load is computation-intensive and how many resources are required to support local execution. Besides network signal strength [21, 40] found that data size and available bandwidth also affect transmission energy consumption. When transmission energy consumption exceeds local computation energy consumption, edge computing is more suitable. If the user focuses on the entire edge computing process rather than the terminal, the total energy consumption should equal the sum of each layer's energy consumption. Like the endpoint layer, other layers' energy consumption equals the sum of local and transmission energy consumption. Workload distribution strategies need to optimize this balance. When the local data center layer is busy, loads need to be uploaded to higher layers. Multilevel data transmission incurs extra overhead, increasing energy consumption compared to executing tasks at the terminal.

5.2.3.4 Cost

From the perspective of service providers (e.g., YouTube, Amazon, and Taobao), edge computing can ensure lower latency and energy consumption, increasing throughput, improving user experience, and ultimately leading to higher profits. For example, based on residents' preferences, a video could be played at the building layer edge, while the city layer edge handles more complex tasks to increase overall throughput. Service providers invest in building and maintaining each layer of devices. To fully utilize local data at each layer, providers can charge users based on data location. Developing new cost models that ensure service provider profits and user affordability is an urgent issue.

Thus, load distribution needs to consider the interrelation between these metrics. For example, due to energy constraints, a workload might need to be completed at the city data center layer, where energy limitations impact latency more than at the building service layer. Optimization metrics should be weighted and prioritized for different workloads to systematically choose a reasonable distribution strategy. Additionally, cost analysis should be conducted during operation, and service providers should consider the interference between concurrent loads and resource usage.

5.3 Security, Privacy, and Service Management

As edge computing continues to expand, security, privacy, and effective service management have become paramount concerns. The distributed nature of edge networks introduces unique vulnerabilities that must be addressed to protect sensitive data and ensure system integrity. In this section, the strategies for safeguarding privacy and security in edge environments will be discussed, as well as the challenges of managing edge services effectively in a decentralized infrastructure.

5.3.1 Privacy Protection and Security

Privacy protection and security are critical services provided by edge computing. For instance, in an Internet of Things (IoT) system deployed within a home, a significant amount of private information is captured by sensors. Providing services while protecting privacy is a challenge. Shi et al. [43] found that performing computations near the data source is an effective method for protecting privacy and data security. The research on privacy protection and security in edge computing faces the following challenges:

Awareness of Privacy and Security in Society: Taking WiFi network security as an example, a survey [12] indicates that among over 400 million homes using wireless connections, 49% of WiFi networks are insecure, and 80% of households still use the default passwords to set up their routers. For public WiFi hotspots, 89% are insecure. If users do not protect their personal privacy data, it is easy for others to use devices like network cameras and health monitors to spy on personal data.

Dual role of edge devices as data collectors and owners: Data collected by devices like smartphones is stored and analyzed by service providers. A better way to protect privacy is to keep data at the edge and let users own their data. Data collected by network edge devices should be stored locally, and users should have the right to restrict service providers' use of this data. To protect user privacy, highly sensitive data should be deleted from edge devices.

Lack of effective tools for data privacy and security: Network edge devices have limited resources, and existing data security methods are not fully applicable to edge computing. The highly dynamic environment at the network edge also makes it more vulnerable to attacks. To enhance the protection of private data, researchers are studying privacy protection platforms. For example, the Open mHealth platform developed by Deborah's team [22] standardizes the processing and storage of health data. However, future research needs to develop more tools to handle data in edge computing.

5.3.1.1 Sensor Security

In today's landscape, edge devices like autonomous vehicles integrate diverse sensors (e.g., cameras, global navigation satellite system (GNSS), and LiDAR) to perceive their surroundings. These sensors face direct security threats, particularly from attacks that manipulate or obstruct sensor data without compromising the computing system itself. Attackers exploit vulnerabilities in sensor operation to interfere, obscure, or falsify data, thereby disrupting edge device functionality [35].

Cameras serve as fundamental visual sensors in intelligent and surveillance systems. Modern edge devices often deploy multiple cameras with various lenses [15, 28]. Camera inputs are pivotal in tasks such as object detection and tracking. For instance, in autonomous vehicles, a popular type of edge device, attackers can deceive these systems by placing counterfeit traffic signals, signs, or objects (e.g., vehicles and pedestrians), leading to incorrect decisions [36]. Attackers also employ high-brightness IR lasers to blind cameras by interfering with infrared wavelengths, thus compromising their ability to provide accurate visual data [36, 46].

Edge devices rely on GNSS and inertial navigation systems (INS) for real-time location updates. GNSS sensors are susceptible to jamming and spoofing attacks, where attackers disrupt receiver function using out-of-band or in-band signals [20]. Furthermore, attackers can deploy GNSS transmitters near vehicles to falsify location data by replicating genuine signals [20]. INS sensors, sensitive to magnetic fields, can be manipulated by powerful magnetic interference, resulting in erroneous orientation readings for the targeted edge devices.

LiDAR technology produces 3D environmental data by measuring distances using laser light pulses. Attackers can deceive LiDAR sensors using absorbent or reflective surfaces, causing them to misidentify obstacles in traffic scenarios [35]. Manipulating the laser pulses can further distort LiDAR data, leading to inaccuracies in object position and distance readings. Ultrasonic sensors and radars, crucial for passive perception and as a final defense for edge devices, have also been vulnerable to spoofing and jamming attacks through specialized signal generators and transmitters [50].

5.3.1.2 Securing Edge Networks and Platforms

Security remains a critical concern in edge computing, prompting several studies to address security challenges across various scenarios. Existing literature categorizes these efforts into two main areas: network security within edge environments and security measures within the operational context of edge computing.

Bhardwaj et al. [4] introduced ShadowNet, a framework that deploys edge functions across distributed infrastructure to monitor IoT traffic and preemptively detect IoT-DDoS attacks. Compared to conventional methods, ShadowNet achieves detection 10 times faster and mitigates 82% of traffic before it reaches the broader Internet, thereby reducing overall security risks. Yi et al. [51] proposed an approach leveraging SDN to enhance edge network security through improved monitoring, intrusion detection, and resource access control, offering valuable insights into mitigating network threats in edge computing environments.

Ning et al. [34] evaluated various trusted execution environments (TEEs) like Intel Software Guard Extensions (SGX), ARM TrustZone, and AMD secure encrypted virtualization (SEV) on heterogeneous edge platforms. Their work demonstrates the deployment of TEEs to bolster security with minimal performance overhead in edge computing scenarios [34]. Additionally, Li et al. [26] developed Kernel Level Resource Auditing (KLRA), a tool tailored for IoT and edge operating systems. KLRA monitors system behavior at a granular level, promptly issuing security alerts upon detecting abnormal system activities, thereby fortifying operating system security on edge devices.

5.3.1.3 Data Sharing Security

Edge computing offers significant advantages through the generation of vast real-time data streams from diverse devices, sites, and infrastructures [39]. Analyzing these data and making informed, multidimensional business decisions can substantially enhance industrial production efficiency [23]. However, traditional edge computing systems tend to be dominated by vertical, closed applications that focus solely on maintaining the operations of individual machines or sites, thereby creating isolated data silos. Integrating edge computing helps break down these data silos and enhances data flexibility. Nevertheless, this integration introduces complexities in securely sharing data, particularly concerning data security and sharing among various specialized applications and stakeholders.

Two primary challenges arise in edge data sharing: firstly, the proliferation of data interfaces increases the risk of severe consequences such as intrusion and data breaches; secondly, the performance limitations of edge devices often hinder the direct implementation of robust security algorithms. The adoption of blockchain technology within edge computing introduces both new challenges and opportunities for securely sharing data [14].

Distributed Edge Data Storage Blockchain-based distributed data storage is pivotal for enabling secure data sharing in edge computing. When integrated with blockchain, data from edge devices becomes tamper-proof, enhancing

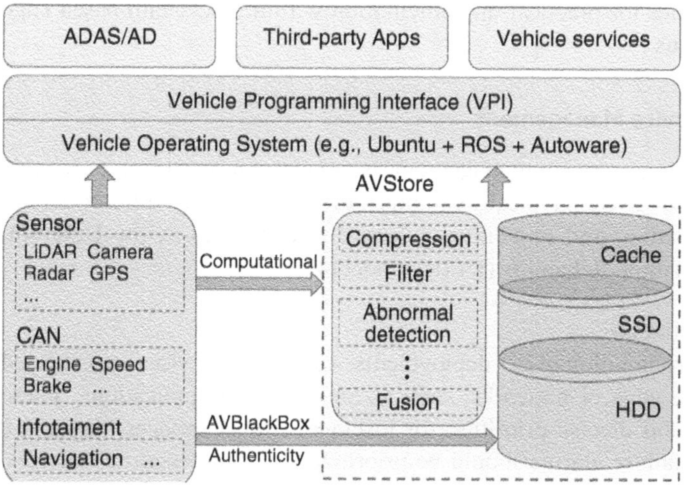

Figure 5.6 Storage system architecture. Source: Wang et al. [47]/ACM, Inc.

security [49]. Recent studies on distributed storage leveraging technologies like the InterPlanetary File System (IPFS) demonstrate its effectiveness as a scalable solution. Storing transaction data in IPFS and including the IPFS hash value in blockchain blocks significantly reduces blockchain data volume. Blockchain also opens avenues for novel business models, such as monetizing edge services. Despite these advantages, there remains a dearth of comprehensive research on systematically integrating blockchain into edge computing to ensure secure data sharing. Addressing these gaps is crucial for advancing the field and resolving outstanding research challenges.

Recently, Wang et al. [47] performed a comprehensive analysis of onboard sensors and controller area network (CAN) bus data in vehicles, which are a type of edge device. They introduced a novel mathematical model that addresses the crucial aspect of data storage requirements for autonomous vehicles, as illustrated in Figure 5.6.

Access Control for Edge Data Secure data sharing hinges on implementing effective access control schemes, which form the foundation of data security in various applications, including connected vehicles [25, 49, 56]. While blockchain integration in edge computing systems can address some data security issues, it introduces challenges such as public data exposure and significant privacy concerns due to data visibility to all system nodes. Therefore, exploring access control solutions that combine attribute-based access control with blockchain holds promise for achieving optimal access control effectiveness. However, the inherent distribution and heterogeneity of edge computing environments necessitate tailored approaches for practical implementation within blockchain-based edge computing systems.

5.3.2 Edge Service Management

In edge computing, service management is crucial for ensuring a reliable system. Any reliable system typically exhibits four characteristics: differentiation, extensibility, isolation, and reliability, collectively referred to as the Different, Extensibility, Isolation, Reliability (DEIR) model [43].

5.3.2.1 Differentiation

With the rapid proliferation of IoT deployments, a variety of services are expected to operate at the edge of networks, such as Smart Home applications. These services will exhibit diverse priorities; for instance, critical services like device diagnostics and failure alarms should be prioritized over nonessential services. Similarly, health-related services such as fall detection or heart failure monitoring should take precedence over entertainment services.

5.3.2.2 Extensibility

Extensibility poses a significant challenge at the edge of networks. Unlike mobile systems, IoT devices are highly dynamic. Can newly purchased devices seamlessly integrate into existing services? Can replacement devices easily adopt the roles of their predecessors? These questions necessitate a flexible and extensible design of the service management layer in edge operating systems (edgeOS).

5.3.2.3 Isolation

Isolation emerges as another critical issue at the edge of networks. In mobile operating systems, an application failure often leads to system crashes and reboots. In distributed systems, shared resources are typically managed using synchronization mechanisms like locks or token rings. However, in smart edgeOS environments, these challenges become more complex. Multiple applications may share the same data resources—for example, light controls. If one application fails, users should still retain control over their lights without system-wide disruption. Similarly, if a user removes the sole application controlling lights, the lights should remain operational rather than lose connectivity to the edgeOS. Addressing these challenges may involve deploying a robust framework for application deployment and undeployment. Detecting conflicts before application installation can warn users and prevent potential access issues. Furthermore, effective isolation must safeguard user privacy by ensuring that third-party applications cannot access sensitive personal data (e.g., activity tracking vs. electricity usage data). Implementing well-designed access control mechanisms within the service management layer of the edgeOS is crucial to resolving these issues effectively.

5.3.2.4 Reliability

Reliability represents a pivotal challenge at the edge of network environments, encompassing perspectives from service, system, and data considerations.

- **Service perspective**: Identifying the precise cause of service failures at the edge can be inherently complex. For instance, when an air conditioner malfunctions, potential causes might range from a severed power cord to compressor failure or depleted battery in the temperature controller. Sensor nodes can easily disconnect due to battery depletion, poor connectivity, or component wear. Merely maintaining current services during node disconnections is insufficient; informing users about nonresponsive components or preemptively alerting them to potential failures would significantly enhance user experience. Adaptations from wireless sensor networks or industrial protocols like PROFINET [11] offer potential solutions to address these challenges.
- **System perspective**: Maintaining the network topology and ensuring each system component can transmit status and diagnostic data to the edgeOS are critical system-level requirements. This capability facilitates tasks such as failure detection, device replacement, and data quality assurance throughout the system.
- **Data perspective**: Reliability challenges in data sensing and communication are prevalent at the edge. Devices may fail due to various factors and can transmit low-fidelity data under unreliable conditions such as low battery levels [6]. New IoT communication protocols have been proposed to support a large number of sensor nodes and dynamic network conditions [9], yet their reliability often falls short of standards set by protocols like Bluetooth or WiFi. Addressing how systems can maintain reliability despite unreliable sensing and communication requires leveraging multiple reference data sources and historical records, posing an ongoing challenge.

Moreover, the concept of a “function cache” has been introduced for managing the lifecycle of edge services [29]. This is particularly crucial when multiple services operate on resource-constrained edge devices, as ensuring that limited

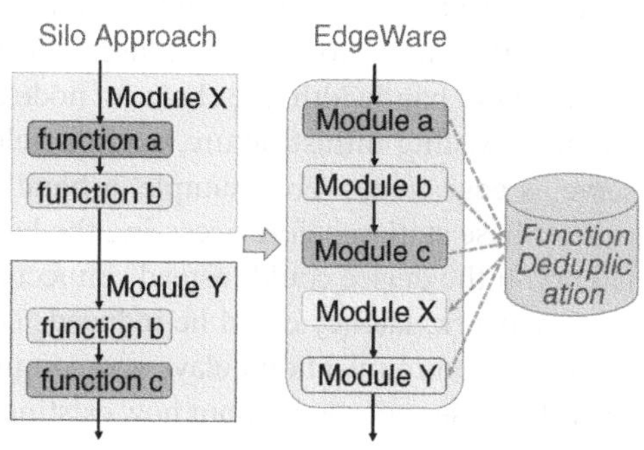

Figure 5.7 An example of function consolidation and deduplication. Each edge service is encapsulated into a function module and the function is consolidated for faster reuse. Source: Lu et al. [29]/Springer Nature.

resources can dynamically support the required services is vital for both automakers and researchers in the field. Efficient and dynamic management of edge services thus becomes essential. As illustrated in Figure 5.7, Lu et al. [29] presented EdgeWare, an open-source, extensible, and flexible middleware designed to manage edge service execution. EdgeWare offers four key features: (1) on-demand model switching, enabling the easy transition and upgrading of machine learning models, (2) function consolidation and deduplication to eliminate redundant copies of recurring functions and maximize the reusability of vehicle services, (3) the creation of event-driven applications to reduce workload, and (4) dynamic workflow customization, allowing for the extension of functionality through customizable workflows.

5.4 Deployment Strategies and Integration

Deploying edge computing systems requires careful consideration of both the hardware and software components, especially when integrating with vertical industries. The unique requirements of deploying edge nodes and AI models on resource-constrained devices pose significant challenges. This section explores the strategies for successful deployment, including the selection of appropriate hardware and software, and discusses how edge computing can be tailored to meet the specific demands of various industrial applications.

5.4.1 Edge Nodes Deployment

The rise of edge computing has attracted significant interest across industries, but the real-world deployment of edge nodes presents several critical challenges that need to be addressed. Key issues include how to effectively select edge nodes and data sources for computation, and how to ensure the reliability of these edge nodes.

5.4.1.1 Selecting Edge Nodes

In practical applications of edge computing, users have the flexibility to choose edge nodes along the path from the cloud to the endpoint to reduce latency and bandwidth usage. However, because edge nodes vary in computational power and network bandwidth, the choice of node can significantly impact computation latency. Existing infrastructure, such as telecommunications base stations, can serve as edge nodes. For example, a handheld device typically connects to the nearest base station before accessing the backbone network, which can increase latency. If the device could instead connect directly to an edge node on the backbone network, latency could be reduced. Thus, selecting the right edge node to minimize communication delays and computational overhead is a critical issue. This also raises questions about how existing infrastructure can be integrated with edge nodes, whether edge computing will foster a new ecosystem, and whether it will bring about revolutionary changes to the current infrastructure.

5.4.1.2 Choosing Edge Data

With numerous edge nodes generating diverse types of data, these datasets often overlap, presenting multiple potential solutions for a given problem. For instance, in real-time traffic monitoring, vehicle speed can be calculated using data from on-board cameras, traffic lights, or roadside units. The challenge lies in selecting the optimal data source for a specific application to minimize latency and bandwidth while maximizing service availability.

5.4.1.3 Ensuring Edge Node Reliability

In edge computing, data storage and computational tasks are heavily dependent on edge nodes, which lack the robust infrastructure of cloud data centers. Many edge nodes are exposed to environmental factors, making their reliability a key concern. For example, public safety solutions that rely on computer vision use smart cameras for data storage and processing. However, these cameras are vulnerable to physical damage under extreme weather conditions, such as strong winds shifting their angles or heavy snowfall obstructing their view. In these scenarios, additional infrastructure is needed to ensure the physical reliability of edge nodes. Moreover, since edge data often has unique spatial and temporal characteristics, it is critical to design effective backup mechanisms to ensure data reliability. Addressing the physical and data reliability of edge nodes through infrastructure support is a critical research area.

Several initiatives have already been undertaken to deploy edge nodes effectively. Many cloud service providers are now offering edge nodes that push high-bandwidth, low-latency, and localized services closer to the network's edge [2, 17, 32]. This approach not only improves service efficiency and capabilities but also maximizes the benefits of multiple stakeholders in the edge computing ecosystem.

5.4.2 Deployment of AI Models on Resource-Constrained Edge Devices

In current edge computing systems, edge devices are typically limited to performing lightweight computing tasks. To enable edge devices and servers to handle more complex tasks with improved data-processing performance and reduced latency, edge intelligence is applied within the edge computing scenarios. This trend represents a significant development in edge computing for industrial IoT (IIoT) environments [39]. However, training and deploying AI models on edge devices pose challenges due to their constrained computing and storage resources. Resolving this conflict involves two fundamental approaches: enhancing the computing power of edge devices and simplifying or partitioning AI models deployed on these devices.

5.4.2.1 Edge Intelligence Devices

Edge devices gather substantial data and require fast, accurate computing models to provide responsive feedback [126]. Deploying AI algorithms on edge devices allows these models to leverage extensive source data for enhanced accuracy, enabling timely and precise decision-making. This integration theoretically achieves a synergistic blend of edge computing and AI [38]. Yet, optimizing devices for edge AI and realizing their full complementarity remains a formidable challenge.

To bolster the computing capabilities of edge devices, intelligent processing modules or specialized AI chips are integrated. These hardware components typically include general computing modules like CPUs, GPUs, and FPGAs, along with customized AI processors tailored to specific device requirements. While general computing modules are widely used for training and inference of AI models on edge devices, customized AI processors are increasingly tailored to specific edge applications and scenarios, with ongoing advancements in technology and architecture.

5.4.2.2 Edge Intelligence Models

Machine learning, particularly deep learning utilizing artificial neural networks [44], represents a potent approach for practical AI applications. However, deploying deep learning models at the edge is hindered by their complexity and computational demands.

(i) **Model Partition**: Deep learning models can be architecturally partitioned for deployment in edge computing environments using three main architectures: independent deployment, collaborative deployment among devices, and device-server collaboration. Independent deployment across multiple devices risks overloading individual edge devices. Collaborative architectures distribute portions of the neural network across multiple devices to optimize performance and resource utilization. Research in this area is nascent, necessitating further exploration and refinement of partitioning strategies [58].

(ii) **Model Size Reduction**: Given the inherent computational limitations of edge devices, model simplification techniques are crucial to enhance processing efficiency. Methods such as weight pruning and data quantization are commonly employed. Weight pruning involves prioritizing neurons based on their contributions and eliminating less-impactful neurons to reduce model size [19]. Data quantization reduces computational overhead by representing model inputs and outputs with fewer bits, thereby accelerating operations.

5.4.3 Integration with Vertical Industries

In cloud computing scenarios, users from various industries can transmit their data to centralized cloud computing centers, where IT professionals handle storage, management, and analysis tasks. This model allows IT professionals to focus on the data itself without needing in-depth knowledge of the user's specific industry.

However, in the context of edge computing, where edge devices are closer to data producers, there is a much tighter relationship with vertical industries. Designing and implementing edge computing systems require significant domain expertise. Vertical industries, eager to leverage edge technologies to enhance their competitiveness, often face a lack of specialized computing expertise. Therefore, IT professionals must collaborate closely with these industries to develop practical, deployable computing systems. This collaboration involves addressing three key challenges: bridging gaps with industry standards, enhancing data protection and access mechanisms, and improving interoperability with existing systems.

5.4.3.1 Bridging the Gap with Industry Standards

Different industries have accumulated years of experience and established standards that must be respected when designing edge computing systems. To minimize gaps, these systems need to align with these industry-specific standards. For instance, in the development of connected and autonomous vehicles, successful implementation requires expertise in intelligent algorithms, embedded systems, and automotive control, which necessitates close cooperation with traditional automotive manufacturers. Similarly, in sectors like manufacturing and industrial IoT, it is crucial to design edge computing systems that adhere to industry standards to ensure successful integration and deployment.

5.4.3.2 Enhancing Data Protection and Access Mechanisms

In edge computing, data is often stored on devices closer to the data source, which provides a level of privacy but also creates challenges in terms of data sharing and access. Industries like healthcare and law enforcement, which deal with highly sensitive information, may be reluctant to upload data to public clouds. Edge computing offers an advantage by keeping data localized, thereby enhancing privacy. However, this also leads to fragmented data storage environments, complicating data sharing and access. It is crucial to develop unified, user-friendly data sharing and access mechanisms that maintain privacy while ensuring the accessibility of data across the necessary platforms.

5.4.3.3 Improving Interoperability

Edge computing systems must be designed to integrate seamlessly with existing industry systems, taking into account the current landscape and existing technologies. For example, in video surveillance systems, although smart cameras with built-in computing capabilities have become more common, there are still many traditional, nonintelligent cameras in use, generating vast amounts of video data daily. Processing this data often involves utilizing existing infrastructure, such as nearby stores or gas stations [57], which may not always have the necessary computing power on-site. To address this, edge computing research should focus on how to deploy edge computing devices effectively in such environments. Current solutions often involve building more data centers or deploying AI-integrated devices, but these approaches can be costly and may revert to a cloud-centric model. Therefore, the challenge lies in developing edge computing systems that are both practical and interoperable with existing technologies, ensuring they can be seamlessly integrated into the industry's current operations.

To address this challenge, the Linux Foundation has provided an open-source initiative, LF Edge [45], a comprehensive ecosystem for the development and deployment of edge computing solutions across multiple industries. It consists of various projects, such as EdgeX Foundry and Akraino edge stack, to offer modular, flexible, and scalable frameworks for integrating edge computing into vertical industries like telecommunications, healthcare, and manufacturing.

EdgeX Foundry is a highly flexible open-source software framework designed to enable interoperability between IoT devices and applications at the edge. EdgeX Foundry provides a modular reference architecture for data ingestion, normalization, analysis, and sharing, which is critical in environments like smart cities, retail, building automation, and manufacturing. It supports a wide range of protocols (e.g., message queuing telemetry transport (MQTT), REST, and Bluetooth low energy (BLE)) and offers enhanced security features, making it a key tool in standardizing IoT frameworks across different market verticals.

Akraino edge stack provides a collection of open-source blueprints tailored for building edge infrastructure across a variety of use cases, such as 5G, AI, and IoT. These blueprints are designed to address the unique challenges of edge environments by focusing on low latency, high availability, and scalability. Akraino supports multiple workload types, including VMs, containers, and microservices, and offers zero-touch provisioning and automated lifecycle management, which are crucial for reducing operational complexity and costs in edge deployments.

By adopting LF Edge, organizations can leverage these tools to address specific deployment challenges, ensuring seamless integration, enhanced security, and reduced complexity in managing edge infrastructure. LF Edge not only accelerates the deployment of edge nodes but also enables the efficient use of AI models on resource-constrained devices, thereby optimizing the overall edge computing architecture within industry-specific contexts.

5.4.4 Hardware and Software Selection

Edge computing systems are characterized by fragmentation and heterogeneity. On the hardware front, a variety of computational units are utilized, including

CPUs, GPUs, FPGAs, and ASICs. Even within the same category of computational units, products can vary significantly in their capabilities. For instance, NVIDIA's edge hardware offerings include the highly capable Drive PX2 and the more modestly powered Jetson TX2.

On the software side, particularly in the domain of deep learning, numerous frameworks such as TensorFlow, PyTorch, are employed. Each combination of hardware and software exhibits unique performance characteristics across different application scenarios.

Taking image classification as an example, there are numerous AI models with varying levels of computational complexity and accuracy. Additionally, the diversity of edge hardware options, each with different computational capacities, further complicates the selection process. The presence of multiple computing frameworks, each performing differently on various edge hardware platforms, adds another layer of complexity. Consequently, identifying the optimal combination of models, frameworks, and hardware involves significant deployment costs and trial-and-error, leading to a considerable challenge: developers often face difficulties in selecting the appropriate hardware and software products to meet their specific application requirements.

In making hardware and software selections, it is essential to conduct a thorough analysis of the computational requirements of the application. This allows for the identification of hardware that meets the necessary computational capacity. Additionally, selecting a suitable software framework for development is crucial while also considering power consumption and cost constraints. Therefore, the design and implementation of tools capable of assisting users in analyzing the performance and power consumption of edge computing platforms and providing informed recommendations for hardware and software selection is extremely important.

5.5 Foundations and Business Models

Understanding the theoretical foundations and business models behind edge computing is essential for grasping its full potential and impact. This section provides an overview of the key theoretical concepts that underpin edge computing, from its architectural principles to its role in distributed systems. Additionally, we examine the emerging business models that are shaping the edge computing market, offering insights into how this technology can drive economic value and transform industries.

5.5.1 Theoretical Foundations

Brewer et al. [5, 13] developed search engines and distributed web caching, which led them to hypothesize about data consistency, service availability, and partition tolerance. He presented this hypothesis at the 2000 PODC conference, and it was later proven and established as the CAP theorem (Consistency, Availability, and Partition Tolerance theorem) [16]. The CAP theorem is a fundamental theory in distributed systems, particularly in distributed storage.

Therefore, in the research of computer systems based on the edge computing model, establishing a theoretical foundation for edge computing will be a critical challenge for both academia and industry. Edge computing is a highly integrative scientific research field, encompassing computing, data communication, storage, and energy optimization. On one hand, edge computing theory can be based on multiobjective optimization theory to achieve comprehensive optimization of computing, data communication, and energy consumption. On the other hand, specific theoretical foundations can be established in different dimensions such as computing, data communication, storage, and energy optimization.

For example, in the computing dimension, load balancing theory for computing tasks can guide the allocation of tasks between cloud centers and edge nodes, maximizing the efficiency of computing resources. Similarly, based on load balancing and distributed system theory, data communication between the edge and the cloud can be optimized to maximize network transmission bandwidth. Research on distributed multidimensional energy consumption models for edge devices (such as using multidimensional Lyapunov theory) can establish energy efficiency models for multiple edge devices, optimizing energy consumption and improving the utilization of limited energy resources. Additionally, reliability theories like the Lyapunov reliability theory can be used to develop edge computing reliability theories based on multiple edge devices.

The theoretical foundation of edge computing is not yet mature. It needs to integrate well-established theoretical foundations from multiple disciplines, including computing, data communication, storage, and energy optimization, to propose comprehensive or multidimensional edge computing theories. Addressing this critical issue is essential for advancing research in edge computing. A sound theoretical foundation will provide significant guidance for academia and industry in future research and development of application services based on the edge computing model.

5.5.2 Business Models

The business model for cloud computing is relatively straightforward. Users purchase services from relevant providers based on their needs. Specifically, the cloud services provided by cloud computing are an extension of Internet-related services, involving the provision of dynamic, scalable, and virtualized resources over the Internet. Clients of cloud computing services can obtain the necessary services on-demand and in a scalable manner via the network. These services can include information technology (IT) infrastructure, software resources, and other Internet-related resources or services. Cloud computing capabilities can also be traded as a service or commodity over the Internet.

Edge computing spans multiple fields, including IT and communication technology (CT), and involves various industry chain roles such as software and hardware platforms, network connectivity, data aggregation, chips, sensors, and industry applications. The business model for edge computing is not just service-driven, where users request specific services, but increasingly data-driven. For example, the Firework model [43] suggests that each user request involves submitting a data request to the data owner (stakeholder), after which the cloud center or edge data owner processes and returns the results to the user. This shifts the traditional unidirectional business model of center-to-user to a multilateral business model of user-to-center and user-to-user.

The business model for edge computing depends on multiple stakeholders involved in the model. A significant issue facing edge computing is how to integrate the existing cloud computing business models to develop a multilateral business model for edge computing.

5.6 Summary and Practice

5.6.1 Summary

This chapter addresses the critical technical challenges and potential solutions in the realm of edge computing, highlighting the necessity for close collaboration among researchers and developers in computer systems, networks, and application services to tackle these issues effectively. The discussed challenges is summarized in Table 5.1.

The chapter begins by discussing programmability, emphasizing the challenges of deploying user applications on heterogeneous edge nodes. It introduces new programming models, such as the Firework model, which enables distributed data processing and computation flows in edge computing environments. The importance of automatic program partitioning is also explained, focusing on distributing tasks efficiently across different edge nodes while considering resources, energy consumption, and response latency.

Next, the chapter addresses the need for a standardized naming mechanism in edge computing to manage device identification and data communication. Solutions like human-friendly naming conventions for smaller, fixed-edge environments are proposed. The complexities of data abstraction are explored, with an emphasis on preprocessing tasks at the gateway level and addressing challenges related to data format diversity, abstraction levels, and data reliability. The chapter also delves into optimizing resource utilization and reducing latency through

Table 5.1 Technical challenges in edge computing.

Challenges and opportunities	Description
Programmability	Challenges in developing and deploying applications on heterogeneous edge nodes, requiring new programming models.
Automatic program partitioning	Partitioning applications efficiently across edge nodes, considering resources, energy consumption, and response latency.
Naming conventions	Developing efficient and standardized naming mechanisms for dynamic and heterogeneous edge environments.
Data abstraction	Preprocessing data at the gateway level, dealing with data format diversity, abstraction levels, and reliability.
Scheduling strategies	Optimizing resource utilization, reducing latency, and enhancing task processing performance in heterogeneous environments.
Data offloading and load balancing	Distributing data and tasks across multiple devices to prevent overload, reduce latency, and improve efficiency and reliability.
Privacy protection and security	Protecting privacy and ensuring security for data, sensors, edge networks, and platforms, including blockchain integration.
Optimization metrics	Using metrics like latency, bandwidth, energy consumption, and cost to optimize load distribution across edge and cloud layers.
Hardware and software selection	Selecting appropriate hardware and software considering computational requirements, performance, and power consumption.
Integration with vertical industries	Aligning edge systems with industry standards, enhancing data protection, and ensuring interoperability with existing systems.
Edge nodes deployment	Selecting optimal edge nodes to minimize communication delays and computational overhead, and ensuring reliability.
Execution of AI models on resource-constrained edge devices	Training and deploying AI models on edge devices with limited computing resources, enhancing performance and reducing latency.
Edge service management	Ensuring reliable and flexible edge service management, handling dynamic and heterogeneous IoT devices and applications.
Theoretical foundations	Establishing robust theoretical foundations for multiobjective optimization in computing, communication, storage, and energy.
Business models	Developing data-driven business models for edge computing, integrating existing cloud models and involving multiple stakeholders.

effective scheduling strategies tailored to the heterogeneous and dynamic nature of edge computing environments.

The necessity of data offloading to balance system load and prevent device overload is covered, with discussions on both full and partial offloading techniques and their respective considerations. Then, the chapter introduces the technical challenges for managing edge services, ensuring a reliable and flexible edge computing system. Privacy protection and security are highlighted as critical issues, including sensor security, securing edge networks and platforms, and data sharing security, with proposed solutions like edge-based data processing and blockchain integration.

Various metrics for optimizing load distribution, such as latency, bandwidth, energy consumption, and cost, are proposed to ensure efficient task allocation across the edge and cloud layers. The chapter also discusses the challenges of selecting appropriate hardware and software for edge computing applications, emphasizing the need for tools to analyze performance and power consumption. It underscores the importance of aligning edge computing systems with industry standards, enhancing data protection, and improving interoperability with existing systems.

Practical challenges of selecting and deploying edge nodes, ensuring their reliability, and integrating them with existing infrastructure are addressed. Techniques for deploying AI models on edge devices are explored, focusing on enhancing computing power and simplifying AI models to overcome resource constraints.

Finally, the chapter concludes by discussing the theoretical foundations and business models of edge computing. It emphasizes the need for a robust theoretical framework to guide future research and development and explores the evolving business models driven by data-centric approaches in edge computing.

5.6.2 Practice Questions

1) How does edge caching contribute to reducing latency in edge computing?
2) Discuss the trade-offs between security and performance in edge computing.
3) What are the challenges in implementing resource management in large-scale edge computing environments?

5.6.3 Course Projects

1) Design a dynamic data management framework that supports real-time data ingestion, processing, and retrieval at the edge. Use open-source tools such as Apache Kafka (https://kafka.apache.org/) for real-time data streaming, and Apache Cassandra (https://cassandra.apache.org) for distributed storage to build the framework. Evaluate the performance in handling large-scale data streams from IoT devices, focusing on latency, throughput, and data consistency.
2) Explore resource allocation strategies in edge-cloud environments and implement a resource allocation mechanism that dynamically balances workloads between edge devices and the cloud, based on real-time conditions like network latency and device capabilities. Use simulation tools such as CloudSim

(https://github.com/Cloudslab/cloudsim) or iFogSim (https://github.com/Cloudslab/iFogSim) to model the edge-cloud environment and implement the resource allocation mechanism. Evaluate the mechanism's effectiveness by running simulations with varying network conditions, task complexities, and resource constraints.

3) Design a privacy-preserving mechanism suitable for an edge computing scenario, such as secure data processing in healthcare. Implement the proposed mechanism using open-source frameworks like EdgeX Foundry or Open Horizon (https://lfedge.org/projects/open-horizon/), focusing on encryption, data anonymization, or secure data transmission. Simulate potential security threats and assess the effectiveness of the proposed mechanism.
4) Explore the integration of edge computing with existing software and hardware solutions in a specific industry (e.g., manufacturing, healthcare, and 5G) and develop a prototype that demonstrates this integration. Use open-source frameworks like EdgeX Foundry (https://lfedge.org/projects/edgex-foundry/) to integrate edge computing with selected software and hardware solutions.

Chapter 5 Suggested Papers

1 Sumit Maheshwari et al. "Scalability and performance evaluation of edge cloud systems for latency constrained applications". In: *2018 IEEE/ACM Symposium on Edge Computing (SEC)*. IEEE. 2018, pp. 286–299.

2 Lanyu Xu, Arun Iyengar, and Weisong Shi. "CHA: A caching framework for home-based voice assistant systems". In: *2020 IEEE/ACM Symposium on Edge Computing (SEC)*. IEEE. 2020, pp. 293–306.

3 Lanyu Xu, Arun Iyengar, and Weisong Shi. "ChatCache: A hierarchical semantic redundancy cache system for conversational services at edge". In: *2021 IEEE 14th International Conference on Cloud Computing (CLOUD)*. IEEE. 2021, pp. 85–95.

References

1 Brian Amento et al. "FocusStack: Orchestrating edge clouds using location-based focus of attention". In: *2016 IEEE/ACM Symposium on Edge Computing (SEC)*. IEEE. 2016, pp. 179–191.

2 AWS. *AWS for the Edge*. https://aws.amazon.com/edge/. Accessed: 2024-08-05. 2024.

3 Tayebeh Bahreini and Daniel Grosu. "Efficient placement of multi-component applications in edge computing systems". In: *Proceedings of the 2nd ACM/IEEE Symposium on Edge Computing*. 2017, pp. 1–11.

4 Ketan Bhardwaj, Joaquin Chung Miranda, and Ada Gavrilovska. "Towards IoT-DDoS prevention using edge computing". In: *USENIX Workshop on Hot Topics in Edge Computing (HotEdge 18)*. 2018.
5 Eric A Brewer. "Towards robust distributed systems". In: *PODC*. Vol. 7. 10.1145. Portland, OR. 2000, pp. 343–477.
6 Jie Cao et al. "A framework for component selection in collaborative sensing application development". In: *10th IEEE International Conference on Collaborative Computing: Networking, Applications and Worksharing*. IEEE. 2014, pp. 104–113.
7 Shiwei Cao et al. "An energy-optimal offloading algorithm of mobile computing based on HetNets". In: *2015 International Conference on Connected Vehicles and Expo (ICCVE)*. IEEE. 2015, pp. 254–258.
8 Byung-Gon Chun et al. "CloneCloud: Elastic execution between mobile device and cloud". In: *Proceedings of the 6th Conference on Computer Systems*. 2011, pp. 301–314.
9 Francis DaCosta and Byron Henderson. *Rethinking the Internet of Things: a scalable approach to connecting everything*. Springer Nature, 2013.
10 Jeffrey Dean and Sanjay Ghemawat. "MapReduce: Simplified data processing on large clusters". In: *Communications of the ACM* 51. 1 (2008), pp. 107–113.
11 Joachim Feld. "PROFINET-scalable factory communication for all applications". In: *IEEE International Workshop on Factory Communication Systems, 2004. Proceedings*. IEEE. 2004, pp. 33–38.
12 Jeanne Ferrante, Karl J Ottenstein, and Joe D Warren. "The program dependence graph and its use in optimization". In: *ACM Transactions on Programming Languages and Systems (TOPLAS)* 9. 3 (1987), pp. 319–349.
13 Armando Fox and Eric A Brewer. "Harvest, yield, and scalable tolerant systems". In: *Proceedings of the 7th Workshop on Hot Topics in Operating Systems*. IEEE. 1999, pp. 174–178.
14 Michael Frey et al. "Security for the industrial IoT: The case for information-centric networking". In: *2019 IEEE 5th World Forum on Internet of Things (WF-IoT)*. IEEE. 2019, pp. 424–429.
15 Andreas Geiger, Philip Lenz, and Raquel Urtasun. "Are we ready for autonomous driving? The KITTI vision benchmark suite". In: *2012 IEEE Conference on Computer Vision and Pattern Recognition*. IEEE. 2012, pp. 3354–3361.
16 Seth Gilbert and Nancy Lynch. "Brewer's conjecture and the feasibility of consistent, available, partition-tolerant web services". In: *ACM Sigact News* 33. 2 (2002), pp. 51–59.
17 Google Cloud. *Google Distributed Cloud*. https://cloud.google.com/distributed-cloud/edge/latest/docs. Accessed: 2024-08-26. 2024.

18 Kiryong Ha and Mahadev Satyanarayanan. "Openstack++ for cloudlet deployment". In: *School of Computer Science Carnegie Mellon University Pittsburgh* 2014 (2015).

19 Song Han et al. "Learning both weights and connections for efficient neural network". In: *Advances in Neural Information Processing Systems* 28 (2015).

20 Rigas Themistoklis Ioannides, Thomas Pany, and Glen Gibbons. "Known vulnerabilities of global navigation satellite systems, status, and potential mitigation techniques". In: *Proceedings of the IEEE* 104. 6 (2016), pp. 1174–1194.

21 Keith R Jackson et al. "Performance analysis of high performance computing applications on the amazon web services cloud". In: *2010 IEEE 2nd International Conference on Cloud Computing Technology and Science*. IEEE. 2010, pp. 159–168.

22 Van Jacobson et al. *Named data networking (NDN) project 2012-2013 annual report*. Tech. rep. Citeseer, 2013.

23 Mohammad Jbair et al. "Industrial cyber physical systems: A survey for control-engineering tools". In: *2018 IEEE Industrial Cyber-Physical Systems (ICPS)*. IEEE. 2018, pp. 270–276.

24 Mohamed Kamoun, Wael Labidi, and Mireille Sarkiss. "Joint resource allocation and offloading strategies in cloud enabled cellular networks". In: *2015 IEEE International Conference on Communications (ICC)*. IEEE. 2015, pp. 5529–5534.

25 Jiawen Kang et al. "Blockchain for secure and efficient data sharing in vehicular edge computing and networks". In: *IEEE Internet of Things Journal* 6. 3 (2018), pp. 4660–4670.

26 Dong Li et al. "KLRA: A kernel level resource auditing tool for IoT operating system security". In: *2018 IEEE/ACM Symposium on Edge Computing (SEC)*. IEEE. 2018, pp. 427–432.

27 Juan Liu et al. "Delay-optimal computation task scheduling for mobile-edge computing systems". In: *2016 IEEE International Symposium on Information Theory (ISIT)*. IEEE. 2016, pp. 1451–1455.

28 Shaoshan Liu et al. "Computer architectures for autonomous driving". In: *Computer* 50. 8 (2017), pp. 18–25.

29 Sidi Lu et al. "EdgeWare: Toward extensible and flexible middleware for connected vehicle services". In: *CCF Transactions on High Performance Computing* 4. 3 (2022), pp. 339–356.

30 Quyuan Luo et al. "Minimizing the delay and cost of computation offloading for vehicular edge computing". In: *IEEE Transactions on Services Computing* 15. 5 (2021), pp. 2897–2909.

31 Pavel Mach and Zdenek Becvar. "Mobile edge computing: A survey on architecture and computation offloading". In: *IEEE Communications Surveys & Tutorials* 19. 3 (2017), pp. 1628–1656.

32 Microsoft Azure. *Azure Stack Edge*. https://azure.microsoft.com/en-us/products/azure-stack/edge. Accessed: 2024-08-05. 2024.

33 Olga Muñoz, Antonio Pascual-Iserte, and Josep Vidal. "Joint allocation of radio and computational resources in wireless application offloading". In: *2013 Future Network & Mobile Summit*. IEEE. 2013, pp. 1–10.

34 Zhenyu Ning et al. "Preliminary study of trusted execution environments on heterogeneous edge platforms". In: *2018 IEEE/ACM Symposium on Edge Computing (SEC)*. IEEE. 2018, pp. 421–426.

35 Jonathan Petit and Steven E Shladover. "Potential cyberattacks on automated vehicles". In: *IEEE Transactions on Intelligent Transportation Systems* 16. 2 (2014), pp. 546–556.

36 Jonathan Petit et al. "Remote attacks on automated vehicles sensors: Experiments on camera and LiDAR". In: *Black Hat Europe* 11. 2015 (2015), p. 995.

37 Deepak Puthal et al. "Secure and sustainable load balancing of edge data centers in fog computing". In: *IEEE Communications Magazine* 56. 5 (2018), pp. 60–65.

38 Tie Qiu et al. "SIGMM: A novel machine learning algorithm for spammer identification in industrial mobile cloud computing". In: *IEEE Transactions on Industrial Informatics* 15. 4 (2018), pp. 2349–2359.

39 Tie Qiu et al. "Edge computing in industrial Internet of Things: Architecture, advances and challenges". In: *IEEE Communications Surveys & Tutorials* 22. 4 (2020), pp. 2462–2488.

40 Dipankar Raychaudhuri, Kiran Nagaraja, and Arun Venkataramani. "MobilityFirst: A robust and trustworthy mobility-centric architecture for the future internet". In: *ACM SIGMOBILE Mobile Computing and Communications* Review 16. 3 (2012), pp. 2–13.

41 Hooman Peiro Sajjad et al. "SpanEdge: Towards unifying stream processing over central and near-the-edge data centers". In: *2016 IEEE/ACM Symposium on Edge Computing (SEC)*. IEEE. 2016, pp. 168–178.

42 Mahadev Satyanarayanan et al. "The case for VM-based cloudlets in mobile computing". In: *IEEE Pervasive Computing* 8. 4 (2009), pp. 14–23.

43 Weisong Shi et al. "Edge computing: Vision and challenges". In: *IEEE Internet of Things Journal* 3. 5 (2016), pp. 637–646.

44 Daniel Svozil, Vladimir Kvasnicka, and Jiri Pospichal. "Introduction to multi-layer feed-forward neural networks". In: *Chemometrics and Intelligent Laboratory Systems* 39. 1 (1997), pp. 43–62.

45 The Linux Foundation. *LF Edge: Building an Open Source Framework for the Edge*. https://lfedge.org/. Accessed: 2024-08-30. 2024.

46 Khai N Truong et al. "Preventing camera recording by designing a capture-resistant environment". In: *UbiComp 2005: Ubiquitous Computing:*

7th International Conference, UbiComp 2005, Tokyo, Japan, September 11–14, 2005. Proceedings 7. Springer. 2005, pp. 73–86.

47 Yuxin Wang et al. "Quantitative analysis of storage requirement for autonomous vehicles". In: *Proceedings of the 16th ACM Workshop on Hot Topics in Storage and File Systems*. 2024, pp. 71–78.

48 Lanyu Xu, Arun Iyengar, and Weisong Shi. "ChatCache: A hierarchical semantic redundancy cache system for conversational services at edge". In: *2021 IEEE 14th International Conference on Cloud Computing (CLOUD)*. IEEE. 2021, pp. 85–95.

49 Kaiping Xue et al. "Fog-aided verifiable privacy preserving access control for latency-sensitive data sharing in vehicular cloud computing". In: *IEEE Network* 32. 3 (2018), pp. 7–13.

50 Chen Yan, Wenyuan Xu, and Jianhao Liu. "Can you trust autonomous vehicles: Contactless attacks against sensors of self-driving vehicle". In: *Def Con* 24. 8 (2016), p. 109.

51 Shanhe Yi, Zhengrui Qin, and Qun Li. "Security and privacy issues of fog computing: A survey". In: *Wireless Algorithms, Systems, and Applications: 10th International Conference, WASA 2015, Qufu, China, August 10–12, 2015, Proceedings 10*. Springer. 2015, pp. 685–695.

52 Shanhe Yi et al. "LAVEA: Latency-aware video analytics on edge computing platform". In: *Proceedings of the 2nd ACM/IEEE Symposium on Edge Computing*. 2017, pp. 1–13.

53 Matei Zaharia et al. "Spark: Cluster computing with working sets". In: *2nd USENIX Workshop on Hot Topics in Cloud Computing (HotCloud 10)*. 2010.

54 Lixia Zhang et al. "Named data networking". In: *ACM SIGCOMM Computer Communication Review* 44. 3 (2014), pp. 66–73.

55 Quan Zhang et al. "Firework: Big data sharing and processing in collaborative edge environment". In: *2016 4th IEEE Workshop on Hot Topics in Web Systems and Technologies (HotWeb)*. IEEE. 2016, pp. 20–25.

56 Quan Zhang et al. "Firework: Data processing and sharing for hybrid cloud-edge analytics". In: *IEEE Transactions on Parallel and Distributed Systems* 29. 9 (2018), pp. 2004–2017.

57 Qingyang Zhang et al. "Edge video analytics for public safety: A review". In: *Proceedings of the IEEE* 107. 8 (2019), pp. 1675–1696.

58 Zhi Zhou et al. "Edge intelligence: Paving the last mile of artificial intelligence with edge computing". In: *Proceedings of the IEEE* 107. 8 (2019), pp. 1738–1762.

6

Future Trends and Emerging Technologies*

As technology evolves, new innovations continue to emerge. The combination of edge computing with emerging technologies and the enhancement of edge computing power in new scenarios have brought significant changes to people's lives. This chapter introduces several computing paradigms that have been proposed in recent years, comparing and analyzing their similarities and differences. Furthermore, it explores the applications of edge computing technology in several typical scenarios and presents some research directions worth further exploration in these areas.

6.1 Edge Computing and New Paradigm

The core concept of edge computing is relatively straightforward: pushing or pulling computing from a device or cloud to an edge. In this case, the computing task will be divided into different entities, that is, device, edge, and cloud, thus forming the computing path [49]. In recent years, some computing paradigms have been proposed and emerged, such as sky computing [31], computing power network [33], and meta computing [5], which have gained prominence in various fields. This chapter primarily focuses on clarifying the relationship between these newly emerged computing paradigms and edge computing, while also describing the research progress of edge computing across various verticals.

6.1.1 Related New Paradigms

In recent years, with the development of edge computing, a series of new computing paradigms have gradually emerged. Following their introduction

*This chapter is contributed by Qingyang Zhang.

Edge Computing: Systems and Applications, First Edition. Lanyu Xu and Weisong Shi.

Companion Website: www.wiley.com/go/xu/computing

in chronological order, some of the more typical ones are cloud-edge-device collaboration, sky computing [31], computing power network [33], and meta computing [5]. Cloud-edge-device collaboration has been mentioned since the inception of edge computing. For instance, in our work Firework [49], the concept of the cloud-edge collaboration was already used to describe this concept. This paradigm is essentially a concrete manifestation of edge computing. Similarly, concepts like edge intelligence, which were mentioned earlier, are also specific applications of edge computing.

Sky computing and meta computing, are new computing paradigms that have been proposed in recent years. These paradigms were developed after observing certain limitations of edge computing, leading to proposed improvements. However, they can still be regarded as extensions of edge computing, with added features. For example, sky computing extends from the cloud's perspective, while meta computing extends from the perspective of cross-service providers and security. Computing power network, primarily proposed to integrate the computational power of both the cloud and the edge, providing a unified, accessible, and seamless computing infrastructure for end devices.

As John McCarthy predicted about the future of computing in 1961, "Computing may someday be organized as a public utility just as the telephone system is a public utility. Each subscriber needs to pay only for the capacity he actually uses, but he has access to all programming languages characteristic of a very large system ...Certain subscribers might offer service to other subscribers ...The computer utility could become the basis of a new and important industry." In Sections 6.1.1.1–6.1.1.3, we will briefly introduce sky computing and meta computing.

6.1.1.1 Sky Computing

Sky computing was proposed by Ion Stoica and Scott Shenker from UC Berkeley in 2021 in their paper "From Cloud Computing to Sky Computing" published at HotOS [31]. Unlike the well-known cloud computing, sky computing is seen as the future of cloud computing. It envisions a sky filled with many clouds and aims to solve the problem of cross-cloud integration, breaking down the barriers between different cloud platforms and maximizing the use of cross-cloud data. Therefore, in sky computing, the authors designed a two-layer architecture consisting of a compatibility layer and an intercloud layer as shown in Figure 6.1.

- **Compatibility layer**: In cloud computing, different cloud platforms may adopt different architectures and provide distinct application programming interfaces (APIs). Similar to how traditional operating systems abstract away the differences in hardware functionality, the compatibility layer in sky computing is designed to mask the differences in cloud platforms' implementations. By abstracting various APIs, it provides a unified interface to facilitate application development. Fortunately, thanks to existing unified interfaces in traditional technologies, such as RESTful APIs for network access, S3 protocols

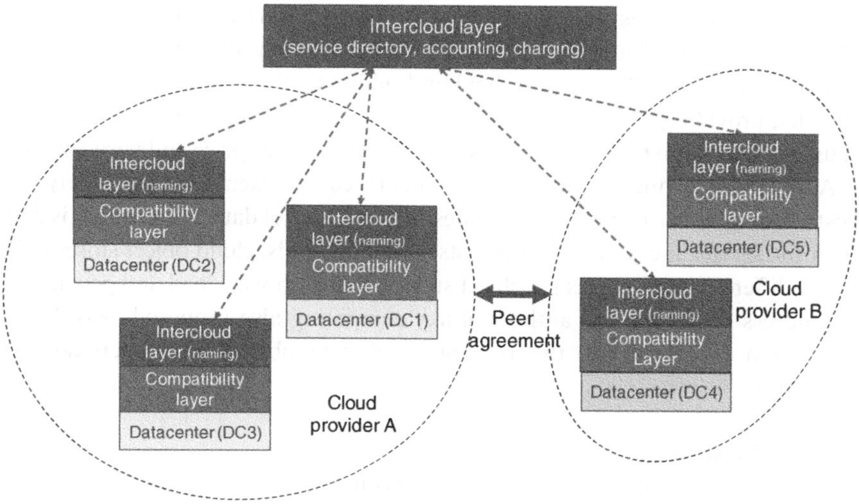

Figure 6.1 Possible sky computing architecture. Source: Stoica and Shenker [31]/ACM, Inc.

for data storage, and Docker or virtual machines for runtime environments, the implementation of the compatibility layer mainly focuses on managing the integration of different cloud service providers' APIs.

- **Intercloud layer**: The intercloud layer is built on top of the compatibility layer and is primarily responsible for providing services to users. It allows users to upload applications and policies. Once the intercloud layer receives a user's request, it will enforce the policy specified by the user to determine where the application should run—whether in a specific location or freely scheduled based on circumstances. Then, the intercloud layer calls the compatibility layer to send requests to the designated cloud service providers for container pulling, launching, and, after the task is complete, for destroying the container or virtual machine. Naturally, the policy can also include additional constraints, such as cost, task completion time, and more.

From the above functionality overview, it is clear that sky computing essentially abstracts away the differences between clouds through the compatibility layer, creating a unified interface for the intercloud layer to invoke. The intercloud layer, in turn, provides an interface for users, allowing them to run applications without needing to know which cloud provider their application is operating on. Users only need to specify constraints, such as runtime, capacity, and cost, and the intercloud layer will automatically find the optimal cloud service provider to deliver the service.

To implement these functions, the sky computing team has proposed several related projects, such as SkyPilot [45] and Skyplane [14]. SkyPilot [45] can be considered the first attempt at sky computing. SkyPilot is a framework for running large language models (LLMs), artificial intelligence (AI), and batch jobs on any cloud, offering maximum cost savings, the highest GPU availability, and managed execution. Skyplane [14], on the other hand, addresses the issue of cross-cloud data transfer, providing a system for bulk data transfers between cloud object stores. It uses cloud-aware network overlays to optimally balance price and performance.

Additionally, while computing can be deployed and executed at relatively low costs, the execution of services generates vast amounts of data. Migrating this data with the service incurs significant costs. In other words, cloud object stores offer vastly different price points for object storage based on workload and geography. To address this issue, the authors of sky computing also proposed SkyPIE [1], a solution for managing the placement of data objects during cross-cloud computing.

6.1.1.2 Computing Power Network

Currently, there is no unified definition for computing power network. The concept was broadly introduced around 2019 as a key infrastructure for computation power, and it can be seen as an evolution of grid computing. The initial goal of a computing power network is to package computation power and deliver it to the places where users need it, much like how electricity is transmitted in a power grid [33, 47]. Although it is still in its early stages of development, different individuals, organizations, etc., provide different explanations for it. Just as with edge computing, most people believe that storing data at the edge and pushing computing power to the edge constitutes edge computing. Similarly, computing power networks have different interpretations depending on perspective.

For example, from the perspective of data centers, the computing power network could be divided into four stages according to maturity: (1) Applications submit tasks and data to a single data center, which then executes them and returns the results. (2) For homogeneous data, applications can be submitted to multiple homogeneous data centers, and the network scheduler eventually dispatches them to a single data center for execution. (3) Applications can run across multiple homogeneous data centers or be submitted to heterogeneous data centers but are ultimately dispatched to a single data center for execution. (4) Similar to the power grid, users only need to submit their application requirements, and the network scheduler coordinates heterogeneous resources to perform the computations. Applications can run simultaneously in heterogeneous data centers and deliver results.

From a broader perspective, a computing power network aims to connect the computing power currently distributed across cloud computing, edge layers, and device layers into a pooled resource using a network. This would allow users to utilize computing power as needed. From this standpoint, computing power networks can be viewed as a further evolution of edge computing, integrating efficient collaboration across cloud, edge, and device layers. However, as previously mentioned, the development of computing power networks is still in its infancy, and its primary drivers are currently large enterprises, particularly those that control data centers and cloud computing centers.

From a broader perspective, a computing power network aims to connect the computing power currently distributed across cloud computing, edge layers, and device layers into a pooled resource using a network. This would allow users to utilize computing power as needed. From this standpoint, computing power networks can be viewed as a further evolution of edge computing, integrating efficient collaboration across cloud, edge, and device layers. However, as previously mentioned, the development of computing power networks is still in its infancy, and its primary drivers are currently large enterprises, particularly those that control data centers and cloud computing centers.

At present, the main focus of computing power networks is still on interconnecting cloud computing, similar to what sky computing is doing. The true integration of computing power across the entire network is still under research and development.

6.1.1.3 Meta Computing

Meta computing is a new computing paradigm proposed by Cheng et al. [5]. As one of the earlier scholars involved in the research of edge computing, Cheng et al. believe that the current computing paradigms are inadequate. Presently, the edge computing environment tends to be isolated, with edge computing nodes still belonging to the same service provider. There is a lack of interoperability and mutual trust between service providers, preventing true collaboration between nodes, as shown in Figure 6.2. Moreover, when there are not many users, maintaining an edge node is actually quite costly, leading to limited resources, service capabilities, and coverage in edge computing.

Although cloud computing and computing power networks have attempted to address some of these issues—such as sky computing, which enables cross-cloud interoperability through the compatibility layer and intercloud layer—these paradigms primarily focus on the cloud side. The edge side is still not fully considered. Meta computing was introduced specifically to solve this problem of cross-domain collaboration. The authors presented a vision for meta computing, aiming to achieve "for any person or task, the entire network functions as a single computer." This is referred to as "network-as-a-computer (NaaC)," and the resulting entity is termed a "meta-computer."

Since computing power barriers across different service providers, security becomes a critical issue. Thus, the authors proposed an architecture design consisting of two core modules as shown in Figure 6.3: the zero trust computing manager module and the device manager module. The device manager module

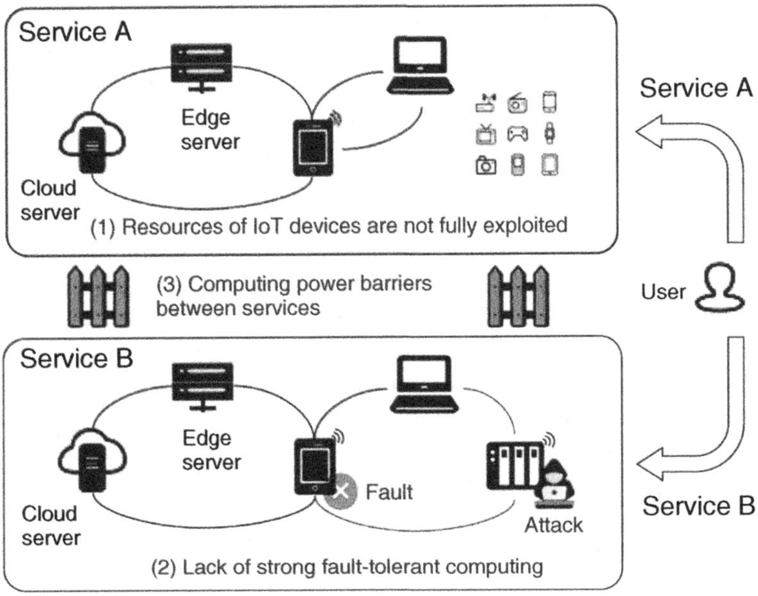

Figure 6.2 Problems of existing computing paradigms. Source: Cheng et al. [5]/IEEE.

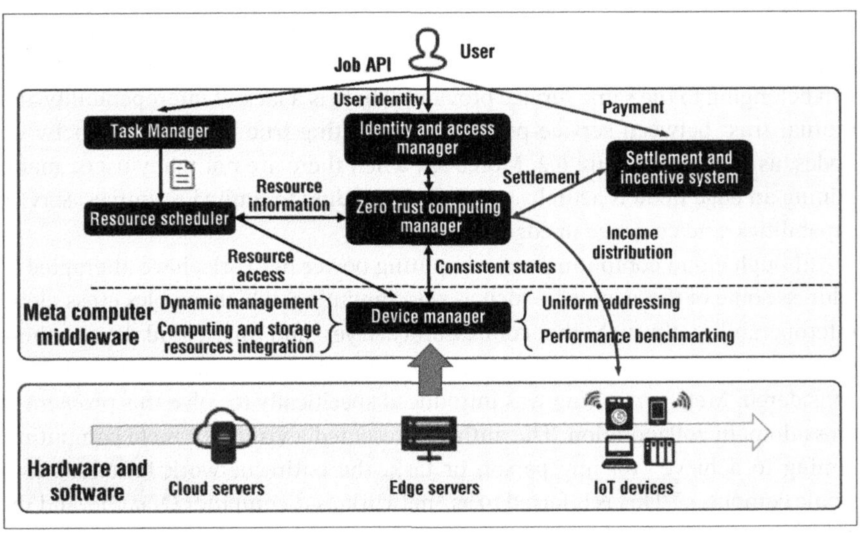

Figure 6.3 Problems of existing computing paradigms. Source: Cheng et al. [5]/IEEE.

Table 6.1 Relationship of different computing paradigms to edge computing.

Name	Features	Relationship to edge computing
Sky computing	Cross-cloud collaboration	Cloud extensions
Computing power network	Data center and edge integration	Cloud and edge extensions
Meta computing	Cross-service provider collaboration	Cross-domain security extensions

aims to unify different computational resources. In the zero trust computing manager module, there are components such as the identity and access manager, the resource scheduler, the task manager, and a settlement and incentive system.

Compared to traditional computing paradigms, the zero-trust-based module is something not found in other computing paradigms. Therefore, we will focus on introducing the functions of these modules. The identity and access manager is used for rigorous identity verification of nodes (including users and devices), thereby creating a zero-trust network and computing environment, while supporting fine-grained management of data access permissions and resource usage. The zero-trust computing manager is responsible for building a distributed ledger, which ensures state consistency in distributed computing tasks through the use of the distributed ledger.

6.1.2 What Is New for Edge Computing

Although some new computing paradigms have been introduced in recent years, they can still be regarded as extensions of edge computing. Table 6.1 summarizes the characteristics of the aforementioned new paradigms and their relationship to edge computing. It should be noted that the edge computing referred to in this book follows the computing path described in edge computing and Firework, involving the device, edge, and cloud layers.

If we only consider their goals, it may seem that only meta computing is related to edge computing, or that only meta computing can be considered an extension of edge computing. However, when we take into account the computing path concept that we proposed earlier, it becomes clear that the cloud is also multifaceted. In Firework, the concept of different holders was mentioned, meaning that the edge and cloud can belong to different service providers. Therefore, if we integrate sky computing and computing power network with edge computing, they both significantly enhance the capabilities of edge computing, as follows:

- **Sky computing**: By connecting different clouds, sky computing enables data from devices to securely and quickly select appropriate compute nodes in a heterogeneous multi-cloud environment, thereby enhancing the collaborative capabilities of edge computing.

- **Computing power network**: Similar to sky computing, computing power networks can achieve comparable functionalities. Additionally, the computing power network plans to incorporate the computing power of the edge layer in the future, effectively merging the edge layer and cloud layer into a unified network. Therefore, the computing power network can be viewed as an enhancement to edge computing from both the cloud and edge layers.
- **Meta computing**: When meta computing was proposed, it directly addressed one of the key issues in current edge computing—cross-domain collaboration. Therefore, when meta computing is fully realized, it effectively becomes a more secure implementation of edge computing. In other words, it can be considered an enhancement to edge computing from a security perspective.

From the above analysis, and as the proposers of meta computing have pointed out, edge computing has shown certain limitations, especially from a security perspective, in terms of its computing paradigm. The current implementation of edge computing is not yet the complete form of what we advocated for in our previous papers. The common practice today is simply to push computing tasks to the edge and label it as edge computing. However, the concept of the computing path has not been fully considered. Particularly, most edge computing solutions are confined to a single service provider, and achieving cross-service provider edge computing remains challenging due to the lack of unified tools.

6.1.3 Future

Whether it is sky computing, computing power network, or meta computing, all have recognized this issue and aim to enhance the computing paradigm from the perspective of heterogeneous, cross-service provider collaboration. Similarly, the realization of our proposed computing path also requires such cross-domain collaboration to fully unlock the potential of edge computing. Therefore, as a major trend for the future, we recommend enhancing cross-domain collaboration capabilities across the cloud, edge, and device layers.

6.2 Integration with Artificial Intelligence

In recent years, with the rise of ChatGPT, large language models have gradually gained attention. Edge computing, on the other hand, enables devices to harness intelligence more effectively, empowering them with greater capabilities. This section discusses the relationship between artificial intelligence and edge computing, highlighting several roles edge computing plays in the training, inference, and caching of large language models. Finally, it explores future applications of edge intelligence and addresses the challenges that currently exist.

6.2.1 Basic Overview and Why Need Edge Computing

A LLM is a type of language model that, compared to traditional models like MobileNet, typically refers to artificial neural networks with billions or even more parameters. Table 6.2 shows the parameter counts of some existing large language models, and these numbers are still growing. LLMs are general-purpose models that perform well across a wide range of tasks, rather than being trained for a single specific task (such as sentiment analysis, named entity recognition, or mathematical reasoning).

It is precisely due to their massive number of parameters that LLMs can capture much of the syntax and semantics of human language and retain a large amount of factual information from their training. The most successful large language model to date is the ChatGPT series developed by OpenAI, which is commonly used for tasks such as text generation, content summarization, sentiment analysis, language translation, and code generation. Additionally, large language models can be customized using techniques like prompting or fine-tuning to generate outputs that better align with user requirements. There are also domain-specific large language models, such as Microsoft's Copilot, which can analyze code and assist in programming based on user requests.

The rise of large models is inseparable from advancements in hardware computing capabilities and the expansion of datasets. At the same time, their training, inference, and other processes require massive amounts of resources. As shown in Table 6.2, the parameter count of GPT (generative pre-trained transformer) models grew from 110 million in GPT-1 to 175 billion in GPT-3.

Table 6.2 Model parameter scales.

Name	R&D team	Parameter scale
GPT-1	OpenAI	0.11 B
GPT-3	OpenAI	175 B
GPT-4	OpenAI	1.8 T
Sora	OpenAI	7 B
Gemini 1.5 pro	Google	175 B
Gemma-2B (opensource)	Google	2 B
Gemma-7B (opensource)	Google	7 B
StableDiffusion	Qualcomm	65 B
LLaMA (opensource)	Meta	7–65 B
ChatGLM (opensource)	Tsinghua University	6–130 B
Hunyuan	Tencent	100 B
ERNIE Bot	Baidu	260 B

After commercialization, the latest GPT-4 expanded tenfold to 1.8 trillion parameters. Training such a massive model required OpenAI to use 25,000 A100 GPUs over 90–100 days, with a single training run potentially costing up to US$ 63 million.

Currently, most large language models are run in the cloud, with only a few capable of operating on high-performance PCs, such as various customized models with around 6 billion parameters. However, to enable better execution on various Internet of Things (IoT) devices or personal portable devices, the models are usually further compressed. For example, Google launched Gemini Nano on the Pixel 8 Pro smartphone, with models having 1.8 billion and 3.25 billion parameters. Qualcomm plans to introduce LLaMA 2 support for flagship smartphones and personal computers powered by Snapdragon chips. Apple also released its own large models in 2024. Apple's OpenELM models come in various sizes, with parameter counts of 270 million, 450 million, 1.1 billion, and 3 billion.

Although these LLMs, with hundreds of millions or billions of parameters, are designed for smart terminals, deploying LLMs on-device remains a significant challenge. However, on-device LLMs can greatly enhance the intelligence of devices, enabling them to provide smarter and more efficient services. Only when all IoT devices exhibit a certain degree of intelligence—whether by becoming intelligent themselves or by collaborating with other devices—can the vision of a fully interconnected intelligent world be realized.

Edge computing plays a critical role in empowering terminal devices by utilizing edge devices and cloud services. It holds great promise in supporting the development of large language models, but there are still many challenges ahead. Specifically:

- Smartphones are not the only part of the Internet of Everything. In the context of the Internet of Everything, devices within the IoT, cyber-physical systems, and similar environments are all connected to the Internet, communicating with each other to complete predetermined tasks. In this scenario, smartphones are just one of the many terminals within the IoE, and they are relatively high-performance devices. However, there are many IoT devices with limited resources, which cannot be equipped with high-performance computing units like smartphones.
- Devices and demands are heterogeneous. The terminal devices in edge computing include a large number of heterogeneous devices. Even among smartphones, their performance varies, and not all smartphones can handle LLMs. In such a heterogeneous environment, collaborative computing in edge computing needs to dynamically adjust based on heterogeneous computing requirements, network demands, latency needs, and more to ensure the availability of LLMs.
- Security and privacy is a big issue. LLMs are trained on vast amounts of data, much of which is user-generated. One of the benefits of edge computing is

that raw data remains local and is only shared after processing, thus mitigating privacy concerns. On-device LLM inference eliminates privacy breaches and reduces the need for Internet connectivity. However, since LLMs require substantial computational power, memory, and energy resources, the challenge lies in how to leverage edge computing technology to accelerate inference and learning of LLMs on the device side.

6.2.2 Integrating LLM with Edge Computing

Edge computing can be seen as a computational paradigm that integrates computing power. When combined with large language models, it can provide benefits in various aspects such as model training, inference, and fine-tuning by leveraging the integration of edge computing.

6.2.2.1 Training with Edge Computing

Training large language models requires vast amounts of data to achieve accurate results. These data may contain significant amounts of user privacy, making on-device learning or fine-tuning the most ideal approach from a privacy perspective. However, large language models have extremely high computational requirements. According to reports, GPT-4 requires approximately 560 trillion floating-point operations for each forward pass to generate a single token. However, advanced A100 GPUs offer only 19.5 trillion floating-point operations per second. This means that with a single A100 GPU, GPT-4 would require around 28 seconds for each forward pass to generate a token. Moreover, backward propagation typically demands even more computational resources than forward passes.

Given that large language models have only been proposed and gained attention in recent years, and since their training typically requires large-scale GPU clusters with tens of thousands of GPUs to achieve optimal results, there has been relatively little research on training models at the edge or device level. This remains an open question and poses a significant challenge. In subsequent text, we will present some hypotheses about the integration of large language models and edge computing, along with some of the work done by scholars in this area.

Centralized Edge Learning In early studies, most researchers referred to computations carried out on edge devices as edge computing. In this case, it only involves offloading the training tasks to edge servers, while devices do not participate in the training process. All data is uploaded to the edge server for training. Currently, some research is exploring this approach for LLM development. For example, Narayanan et al. [25] have conducted studies in this area. They leverage distributed GPUs at the edge to train LLMs. However, this approach is not applicable to most edge nodes. Currently, research in this field remains relatively scarce due to the high resource demands of LLM training, which a single edge device is unable to support. We also note that fine-tuning consumes less than training the entire model, so might we not consider placing the fine-tuning at the edges to construct models that are more suited to edge-specific environments? This is an open-ended question.

Distributed Edge Learning This approach leverages the distributed nature of edge computing, coupling the model learning process with distributed computation, allowing different devices to train the same model. In this setup, federated learning and split learning can be integrated with edge computing in LLM training. The main difference between the two is that in federated learning, the model is not split, meaning all participants train the same model, leading to larger parameter transmission. In split learning, the large language model is divided into smaller submodels, and different devices focus on training specific parts of the model. Eventually, these submodels are aggregated into a single model in the cloud. Compared to centralized edge learning, research on distributed edge learning is more extensive. This is mainly because the resource demands of large language model training are too high for most edge nodes to handle the significant computational load.

In the field of federated edge learning, Wang et al. [38] proposed a cloud-device collaborative learning framework for multimodal large language models. This framework employs a token sampling strategy to filter out some irrelevant tokens, reducing transmission costs and improving training efficiency. It also uses adapter-based knowledge distillation to distill the large language model into smaller models that are easier to deploy on edge devices. During model downloading or updating, it adopts a dynamic weight update compression strategy to reduce transmission costs and balance the differences between cloud and edge models. This framework can also be applied in cloud-edge collaborative environments.

Furthermore, some work places the training in the cloud while allowing fine-tuning to take place at the edge or on devices, which is a promising approach. Additionally, in the context of split learning, Lin et al. [20] proposed an efficient parallel split learning scheme. Looking ahead, particularly in combination with large models, split learning may represent a more viable training paradigm because it enables training sub-models on devices and integrates the complete model in the cloud.

6.2.2.2 Inferring with Edge Computing

Compared to training, the inference of LLMs consumes slightly fewer computational resources [42]. However, it is still difficult to directly apply LLMs to end devices without leveraging edge computing for support. For the LLMs themselves, numerous techniques have been developed to reduce the costs associated with computation and communication. Common compression techniques such as pruning, quantization, and knowledge distillation are frequently employed. For example, through INT4 quantization, Google's Gemini Nano models, with 1.8 billion and 3.25 billion parameters, can run on the Pixel 8 Pro. Similarly, Gemma 2B, after INT4 quantization, can be deployed on iOS devices. Apple also released its LLMs in 2024. The OpenELM models from Apple are available in various sizes, with parameter counts of 270 million, 450 million, 1.1 billion, and 3 billion.

Although these quantization techniques help, LLMs still struggle to be deployed on other edge devices due to the around 10 GB memory requirements and second-level latency involved in computations. Compared to the massive cloud-based models with hundreds of billions of parameters, edge models are more limited in functionality. For example, edge models can only support relatively "basic" functions such as text summarization, suggesting intelligent replies based on context, and checking grammar. Therefore, introducing edge computing to enhance the inference of edge models is crucial to overcome these limitations [12].

Optimizations of LLMs Large language models can be optimized using a number of unique techniques. Among these, the more typical ones include speculative decoding, and early exit.

- **Speculative decoding**: Large models are slow in inference, but while smaller models may have inferior inference capabilities, they can still provide a passable result. Therefore, some have compressed large models into smaller models. These smaller models are then run on edge devices to first obtain a result. The data is subsequently sent to cloud or edge services to execute the complete large model for a more accurate result. Finally, the best result is used to further refine the output of the edge device's model. For example, Wang et al. [37] proposed Tabi, a multistage inference engine system that uses smaller models along with optional LLMs to provide query services for demanding applications. Tabi uses calibrated confidence scores to decide whether to return the smaller model's accurate results at high speed or to reroute them to the LLM. For rerouted queries, it employs attention-based token pruning and weighted ensemble techniques to offset system overhead and accuracy loss.
- **Early exit**: Early exit techniques have been widely used in LLM inferencing to reduce latency. In this case, the model on the end-device or edge-device can be executed only partially instead of having to be executed in its entirety, which can reduce the cost. For example, Chen et al. [4] proposed a framework for large language modeling called early-exit (EE)-LLM with support for both training and early exiting of inference, which improves the overall model training and inferencing speed. At the same time, it can be combined with others to further reduce the overall latency.

Splitting Model with Edge Computing From the inception of edge computing, extensive research has focused on empowering edge devices with artificial intelligence. Researchers have explored methods such as splitting models or processing the entire model on edge devices. The latter approach, where the entire model is processed on the edge device, is more straightforward. However, the former approach—splitting the model for execution—tends to be far more efficient than executing the entire model on edge devices.

One classic work, such as that by Kang et al. [17], involves splitting networks like AlexNet layer by layer. Their research demonstrated the data sizes of the intermediate outputs after splitting and showed that utilizing cloud-edge-device collaboration can effectively reduce overall latency. Similarly, the work of Zhang et al. [48, 49] from both the application and model levels has further confirmed this approach.

In the context of LLMs, the approach of model splitting and deploying along the edge computing path is particularly well suited for LLM inference. On the one hand, compared to on-device LLM inference, split LLM inference offloads most of the computation to edge servers, thereby reducing the workload on the device, which is crucial for computation-intensive LLM inference. On the other hand, considering that LLM applications in edge networks (such as healthcare and autonomous driving) often involve highly sensitive personal data, split LLM inference can effectively alleviate privacy concerns, as the edge device does not need to share private raw data with the edge server.

Following the approach of Kang et al. [17], splitting LLM inference can still be effective. The initial layers of the model can be processed on the device, while the remaining layers are offloaded to the edge server for inference. Some works have already adopted this approach, such as Ohta and Nishio [26] and Ma et al. [23]. Of course, this method still involves significant overhead, placing high demands on the inference capabilities of the edge servers.

Also, edge computing could utilize the feature of LLMs to further reduce the latency in terms of computing and transmitting. For example, some works target token compression in LLMs, which they call token representation reduction. In fact, the core idea of their work is to take the intermediate output layers that need to be segmented, and compress them using quantization, pruning, and so on. For example, Cao et al. [3] inserts a binarization module after layer norm layers to quantize the token representations with 1-bit vectors.

6.2.2.3 Model Caching with Edge Computing

In fact, beyond simply improving model inference and training, model caching is another common approach used to enhance efficiency in distributed inference or learning. Just as edge computing originally evolved from data caching in content delivery networks (CDNs), it can be seen as caching various computing services at the edge. Model caching is also applicable to large models.

In LLMs, techniques such as fine-tuning often allow modifications to the base model, resulting in multiple models that are fine-tuned from the same base. However, these models often share many identical sub-models. By utilizing caching techniques to store these submodels at the edge, both learning and inference can significantly reduce memory overhead on edge devices. Furthermore, batch processing can be employed to accelerate inference. This technique proves especially effective in distributed edge learning, particularly in scenarios like split learning or split inference.

In terms of caching, some studies are already underway. For example, Qu et al. [28] proposed the TrimCaching framework, which focuses on parameter-sharing edge caching for AI model downloading. However, this concept can also be applied to collaborative inference.

6.2.3 Integration with Generative AI

Generative AI refers to a subset of AI that focuses on creating new content [43], such as text, images, music, or even code, by learning patterns from existing data. Unlike traditional Ai models that are designed to recognize patterns or make decisions based on input data, generative AI models can produce original content that wasn't explicitly programmed. This is achieved through techniques such as generative adversarial networks (GANs), variational autoencoders (VAEs), and more recently, LLMs like GPT.

These models have gained significant attention due to their ability to generate high-quality, contextually relevant outputs that can mimic human-like creativity and decision-making. This ability to generate new data or content has broad applications across various fields, including natural language processing, computer vision, and even scientific research [29].

Generative AI can be effectively integrated with edge computing to enhance the capability of IoT systems and other edge applications in a few ways.

- **Real-time Data Processing and Decision-Making**: Generative AI models deployed on edge devices can analyze and generate data in real time, allowing for immediate responses to be made directly on the device. For example, in autonomous vehicles, generative AI could process data from sensors and cameras to predict and react to traffic conditions without needing to communicate with a central cloud server. This capability is critical in situations where low latency and real-time decision-making are essential.
- **Reducing Latency and Bandwidth Requirements**: One of the primary benefits of integrating generative AI with edge computing is the reduction in data that needs to be sent to and from the cloud. By processing and generating data locally, edge devices can reduce the amount of bandwidth required for communication, which is particularly important in environments with limited

connectivity or where data privacy is a concern. For instance, in smart cities, generative AI could be used to process video feeds from traffic cameras to optimize traffic flow in real-time, reducing the need to transmit large video files to a central server.

- **Optimizing AI Models for Edge Devices**: Edge devices typically have limited computational and memory resources, making it challenging to deploy large generative AI models. However, model compression techniques discussed in Chapter 4, such as pruning and quantization, can be used to reduce the size and complexity of these models, making them more suitable for edge deployment. For example, a quantized version of a generative AI model could run on a smartphone or IoT device, generating useful outputs while consuming less power and memory.
- **Enhancing Privacy and Security**: Generative AI allows edge devices to be more powerful when processing data locally and, therefore, enhances privacy and security. Sensitive data, such as personal health information, can be analyzed and acted upon directly on the device, such as personal health information, can be analyzed and acted up directly on the device, without needing to transmit it to a cloud server where it could be vulnerable to breaches.
- **Hybrid Edge-Cloud Architectures**: In scenarios where edge devices are not capable of handling the full computational load required by generative AI models, a hybrid approach can be used. In this setup, edge devices perform lightweight processing, while more computationally intensive tasks are offloaded to nearby edge servers or the cloud. This allows for efficient processing while maintaining the benefits of low latency and improved privacy. For instance, a smart home system might use generative AI to process voice commands locally, while more complex language processing is handled by a cloud server.

Integrating generative AI with edge computing is promising in creating more intelligent, responsive, and secure systems across various industries, This synergy is likely to drive significant advancements in the way we interact with and benefit from technology in the coming years.

6.2.4 Applications and Future

Currently, many applications are being developed based on large models. The most common use cases remain conversational chatbots, followed by various assistants, such as Google Assistant on smartphones. Of course, by enhancing large models with domain-specific knowledge and performing appropriate fine-tuning, we can obtain domain-specific large models that lead to even more applications. For instance, large models can be used in industrial design, medical assistance, and many other fields. These fine-tuned domain-specific large models also give rise to models further optimized through various fine-tuning or prompt engineering, which in turn fosters the creation of a large model marketplace [30]. Next, we will explore several potential large model application domains that could be highly beneficial and valuable in the future.

- **Robots**: By injecting the intelligence of LLMs into robots, significant potential can be unlocked [39], such as achieving humanoids similar to those in the game Detroit: Become Human. Leveraging the powerful capabilities of LLMs, humanoid robots can efficiently perform a wide range of tasks, from assisting in warehouses to executing rescue missions, and providing support in hospitals, elderly communities, and households. In this area, many companies are already actively developing solutions. For example, Tesla is developing the second generation of its general-purpose robot, Optimus, while NVIDIA has its GR00T project. In the academic world, many efforts are also being made in this direction. For instance, Joublin et al. [15] proposed the CoPAL architecture, and Zhao et al. [51] introduced the RoCo system, both of which utilize large language models to achieve collaborative path planning for multiple robots, thus improving task execution efficiency.
- **Autonomous driving**: Most existing autonomous driving solutions rely on a modular approach, dividing driving tasks into independent components such as perception, prediction, and planning. However, modular design inherently has a limited capacity for tasks that require complex, human-like reasoning. And LLMs are suitable for this area while it could be a black box to achieve end-to-end learning and inference [7]. By embedding large language models, not only can better-driving decisions be made [44], but in-vehicle assistant functions can also be provided simultaneously. Additionally, the in-vehicle assistant could anticipate users' needs and adjust driving decisions accordingly, offering a more comfortable driving experience.

Of course, new technologies also face numerous challenges, and this is equally true for large language models. For large language models integrated with edge computing, there is still a long way to go. Several open questions need to be addressed, as listed below.

- **Explainability**: In traditional artificial intelligence, explainability has become a key research focus. The same applies to LLMs. However, due to the more complex architecture of LLMs, how to achieve the feature of explainability is a major issue. At the same time, explainability during processes like model splitting and submodel extraction is a critical consideration in edge computing.
- **Verifiability**: The inferring overhead of LLMs is relatively high, and edge devices may cut corners by not returning the correct results to conserve their

own resources. This is particularly concerning when a small model runs on the endpoint while a large model runs on the edge or in the cloud. The edge service might entirely agree with the results from the small model at the endpoint and skip executing the large model. Although there are already some verifiable execution efforts based on secure multiparty computation, the overhead is too high to be practical in real-world scenarios.

- **Trimmability**: Considering the use of edge caching and other techniques to cache large models and accelerate inference speeds, one could also explore selectively executing only certain modules based on the problem at hand. For lower-impact models, smaller models could be used as substitutes. By trimming the execution based on the problem, the inference speed can be further increased.

6.3 6G and Edge Computing

In the realm of network communications, edge computing is often employed to optimize the quality of service at the network edge. For instance, in the 5G era, by deploying computing resources near base stations, some data can be processed and forwarded at the network edge without needing to route through the core network, thus speeding up data transmission within the network. Similarly, in 6G, recognizing the advantages of emerging technologies like edge computing, there are plans to deploy artificial intelligence at base stations through edge computing, further enhancing network service quality and improving data forwarding efficiency.

6.3.1 Basic Understanding for 6G

6G technology is aimed at communication systems beyond 2030, with its primary characteristic being the integration of artificial intelligence to optimize network communication, ranging from physical channels to data packet forwarding [36]. 6G may also enable ubiquitous AI support, providing devices with embedded AI capabilities [18]. While the specific frequencies for 6G are yet to be determined, the IEEE has indicated that frequencies ranging from 100 GHz to 3 THz are likely candidates for 6G [34]. Some of the typical features of 6G include the following:

- **Increased bandwidth**: With the ongoing advancements in radio interface modulation, coding techniques, and physical layer technologies, the speed of 6G networks is expected to increase significantly. While the peak rate for 5G is 20 Gbps, 6G could achieve peak rates of 1–10 Tbps due to the use of terahertz and optical frequency bands.
- **Diversified communication access**: In 1G through 5G networks, all devices accessed the network via base stations. However, the 6G vision includes the addition of visible light modulation technology, which can enhance signal quality in specific scenarios. Additionally, 6G plans to integrate satellites at various

orbital heights into a unified space-ground network, providing diverse access methods for remote areas, maritime users, and low-altitude unmanned devices, thereby improving network coverage.

- **AI enablement**: As 6G base stations will have limited communication ranges, more base stations will be deployed. With enhanced chip technology, these base stations will also be capable of handling greater computational loads, thereby further enhancing edge computing capabilities. This allows for more AI-driven functionalities, such as intelligent operations and environmental awareness. For instance, the base station could adjust signal transmission frequencies and directions based on user location sensing, providing higher-quality network service (a technology also often discussed in the integrated sensing and communication field [11, 21]). Additionally, base stations may support more edge services.

6.3.2 Mutual Influence: 6G and Edge Computing

Considering the scope of this book, our focus on 6G primarily revolves around its functionalities and the technologies related to edge computing and artificial intelligence. Currently, the roles of these two technologies in 6G are widely recognized. For example, organizations such as 3GPP, IEEE, ETSI, and ITU have all proposed integrating edge computing to achieve 6G in their respective standardization of 6G network architecture/frameworks.

Edge computing, in this context, operates on two levels. At the infrastructure layer, base stations can provide greater computational power, supporting both their own operations and AI-driven wireless communication optimization. Additionally, due to the flexibility of edge computing, it can better coordinate the functions of different base stations and offer personalized AI services. At the application layer, more computational power can be allocated to edge services, enabling a broader range of edge services.

Moreover, the relatively interference-free nature of 100 GHz to 3 THz signals, combined with the smaller coverage area of individual base stations and the lower number of users that need to be served, allows users to experience higher bandwidth and lower latency. These factors collectively enhance the enabling role of edge computing. The strengthening of edge services, in turn, positively impacts the 6G user experience, creating a synergistic relationship that fosters further development. In Sections 6.3.2.1 and 6.3.2.2, we will elaborate on the influence of edge computing on 6G and the opportunities that 6G presents for edge computing.

6.3.2.1 Edge Computing-Enabled 6G

As aforementioned, edge computing technology primarily serves as an enabler of foundational computational power in 6G scenarios. Current research indicates that, particularly during the 5G era, some studies have already explored edge computing-enabled 5G networks, utilizing network slicing optimization to enhance service quality. In 6G, this optimization of network slicing is further enhanced, gradually evolving into edge intelligence-enabled network slicing optimization. Additionally, 6G introduces the ability to sense user location and even user posture through communication signals, a capability made possible by edge intelligence. We will discuss these two topics, focusing on the role of edge computing, particularly edge intelligence, in these areas.

Network Slicing Optimization Network slicing technology is regarded as a key enabler for service optimization in 6G systems. It allows multiple virtual subnetworks to be created on a single physical communication infrastructure, each tailored for different quality of service (QoS) requirements, thereby enhancing the user experience. This technology acts as the role of a traffic controller, directing certain types of data traffic to take the high-speed route directly to the core network while other data traffic takes a regular route. It is clear that the traffic controller must be positioned at the beginning of the network for optimal efficiency, which is why network slicing naturally benefits from edge computing technology. Moreover, deploying caching resources at base stations can further accelerate data retrieval for wireless network users, thereby improving service quality.

Ye et al. [46] proposed a network slicing optimization architecture for 6G systems, enabled by mobile edge computing (MEC), as shown in Figure 6.4. In this architecture, the access point (AP) is connected to a dedicated edge distributed unit, enabling the decentralization of edge computing capabilities. The user-centric distributed unit primarily provides data caching functions and some collaborative computing capabilities.

In their architecture, edge computing is utilized for the joint allocation of communication, computation, and caching resources at different granularities to meet the demands of various latency-sensitive applications. The approach first uses conventional optimization algorithms to determine the optimal allocation and placement of computing resources, followed by the application of multiagent

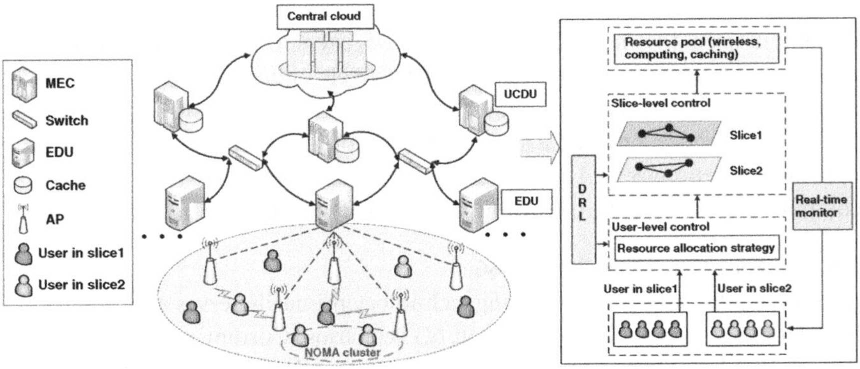

Figure 6.4 A hierarchical network slicing architecture. Source: Ye et al. [46]/IEEE.

deep reinforcement learning to solve problems and strategies that traditional optimization algorithms may struggle to address efficiently.

Reconfigurable Intelligent Surfaces The core idea behind reconfigurable intelligent surfaces (RIS) is to use passive elements to manipulate the scattering properties of 6G electromagnetic waves [32], thereby enhancing the quality of wireless communication. This technology is considered a key communication technology in 6G. Due to its controllable nature—specifically, the ability to enhance signals at certain locations—RIS can be used to provide precise wireless services to individual users in 6G networks. However, delivering such precise services is no easy task. First, the base station must know the user's location; only then can it control its elements to facilitate communication. In this process, artificial intelligence methods are needed to process weak wireless communication signals to determine the user's location, followed by algorithms that adjust the RIS accordingly. As previously mentioned, edge computing plays a crucial role by first enabling the base station with sufficient computational power and then integrating with AI algorithms. Tang et al. [32] and Mukherjee et al. [24] have both made preliminary attempts at the above process in their work. Their results indicate that the use of edge computing can reduce the latency associated with RIS adjustments, thereby improving service quality.

Of course, aside from the aforementioned features, there are many other 6G functionalities that can benefit from edge computing. Overall, edge computing can be seen as providing a foundational layer of computational power to 6G base stations. For example, Bell Labs has proposed a new network architecture concept of radio access network (RAN)-Core integration [35], which unifies parts of the RAN architecture with parts of the core network into a single entity. This approach reduces network complexity and enhances the scalability of network elements and base stations. In this example, the base station can form a cloud-edge collaborative system with the core network, providing greater computational power that supports more AI applications, thereby enhancing the AI-enabled capabilities of 6G.

6.3.2.2 6G-Supported Edge Computing

In Section 6.3.2.1, we mainly discussed how edge computing supports the computational power requirements of the 6G network architecture itself. In this section, we will focus on some applications that integrate edge computing in 6G scenarios. However, since 6G is still in the research phase, actual applications are not yet available. Most of the research is still theoretical, focusing on issues such as the placement of various computational resources. Since this technology has been extensively discussed in Chapter 3, this section will provide a brief introduction, primarily highlighting specific works as examples.

Given the shorter communication distances of 6G, deploying a one-to-one edge server for every 6G base station would be prohibitively expensive. Considering also the reduced cost of data migration in 6G networks (due to higher data transmission speeds), Cong et al. [6] proposed the EdgeGo resource sharing framework for 6G edge computing. In their architecture, edge servers are divided into stationary and mobile types. Stationary servers handle strict real-time tasks, while mobile servers can be used to process non-real-time data or temporarily enhance the capabilities of stationary servers. Thanks to the presence of mobile edge servers, EdgeGo decouples task offloading and execution, allowing mobile edge servers not to remain fixed in one location upon receiving tasks. Instead, they can continue moving and manage the processing results from another transmission path, significantly enhancing mobility and the framework's flexibility, benefiting from the inherent flexibility of 6G networks. Finally, EdgeGo integrates a two-tier iterative optimization algorithm to coordinate optimal solutions for transmission paths and computing task offloading.

Similarly, Huang et al. [13] have conducted research on task scheduling for real-time applications enabled by 6G in edge computing environments. Their work explores task scheduling under the new 6G network architecture. As illustrated in Figure 6.5, which represents their proposed architecture, their scheduling algorithm primarily relies on reinforcement learning to quickly solve Markov decision processes. To ensure the accuracy of the scheduling, they adopt edge learning, where the training of the reinforcement learning model is carried out on edge servers, and the inference process is conducted on the devices.

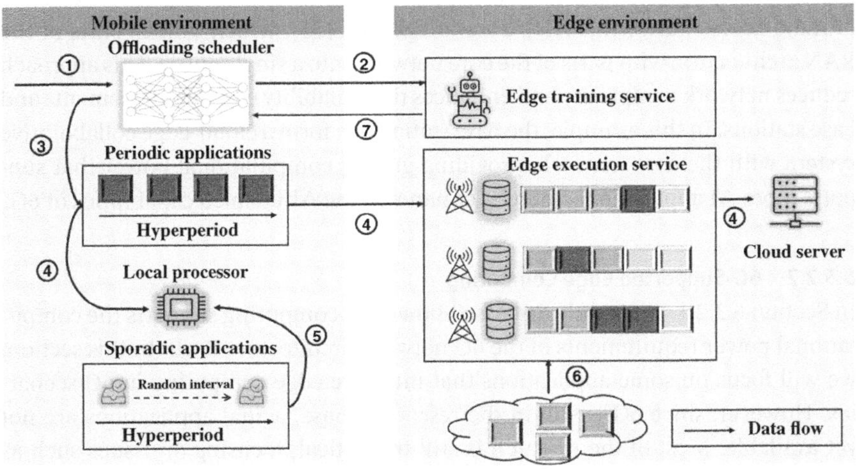

Figure 6.5 Architecture of multi-access edge learning-based offloading (MELO). The data that flow in the MEC system include: (1) environment states; (2) training samples; (3) offloading policy; (4) periodic jobs; (5) sporadic jobs; (6) periodic jobs from other mobile devices; and (7) parameters of edge actor network. Source: Huang et al. [13]/IEEE.

Based on the work discussed above, it is not difficult to see that since 6G networks are still in the pre-research stage, many system-level tasks have yet to be initiated. The primary focus at this point is on task offloading and scheduling. However, whether from the perspective of standards or expert opinions, edge computing is certain to be one of the key enabling technologies in the 6G era, playing a pivotal role and demonstrating its potential in 6G applications.

6.3.3 Potential Applications and Challenges

6G is considered to have potential applications across various fields due to its higher bandwidth and lower latency.

- **Cloud-based virtual reality**: Cloud-based virtual reality (VR) is often mentioned as a key application scenario for 6G in the future. Virtual reality technology requires the transmission of vast amounts of data, such as rendering data, texture data, and more. While some manufacturers have already achieved cloud-based virtual reality under 5G, its functionality remains limited and the related technology has not yet been fully realized. 6G, on the other hand, can provide greater bandwidth on the terminal side than 5G, allowing for faster delivery of various types of data, thereby enhancing the user experience. Additionally, cloud computing enables the sharing of computational power, reducing the computing load on terminal devices. This, in turn, lowers the performance requirements for these devices, allowing for lighter and more portable terminals. Similarly, virtual reality can also be used to deliver holographic video, empowering vertical scenarios like smart healthcare. Of course, edge computing can also bring benefits to cloud-based virtual reality technologies. For example, abundant edge computing resources can be leveraged to provide localized cloud-based VR services, delivering lower latency and further improving the user experience.
- **Communication sensing and digital twin**: The unique channels in 6G enable electromagnetic wave-based positioning and motion sensing of the surrounding environment, a concept previously mentioned as the integration of communication and sensing. As precision sensing technologies mature, the next step is to conduct sensing and modeling of specific scenarios as needed, without the need for additional equipment. For instance, 6G technology can be employed in digital factories for modeling, intelligently sensing various machines and manufacturing processes, and, when combined with digital twin technology, achieving a digitalized smart factory. On one hand, this can be used to predict potential faults and accidents in advance. On the other hand, it can also support technological upgrades like Industry 5.0. Additionally, precise digital twins are considered to be the foundation for technologies like flexible manufacturing.

Although 6G holds great potential for a wide range of applications, from a practical standpoint, it will still take considerable time before final standards are established and large-scale implementation is achieved. Similar to many frontier technologies, 6G is currently still in the pre-research phase, meaning that there are few usable systems available for testing and research. However, we can take advantage of this pre-research period to envision the future and explore potential implementations in 6G. Specifically, there are still many open questions surrounding 6G and edge computing that need to be addressed.

- While intelligent, self-evolving 6G is undoubtedly the trend, the question of how to implement it remains. Specifically, how to deploy significant computational power at base stations, coordinate it across the entire network, and address future potential applications are all factors that will influence this trend. From a systems perspective, how to build a robust, flexible, and scalable architecture, and incorporate it into the relevant standards, in order to provide a degree of flexibility during the initial stages of 6G—enabling it to adapt to rapidly changing future applications—remains an open question. This presents significant challenges for system research.
- While the integration of communication and sensing has inspired many applications, the technology has a dual nature. Sensing the surrounding environment inevitably raises concerns about user privacy. Therefore, another open question is how to address privacy protection at the technical level in order to advance communication-sensing integrated technologies.

6.4 Edge Computing in Space Exploration

With the development of various nanosatellite technologies in recent years, orbital edge computing (OEC) has gradually emerged as a means to enhance the flexibility of satellite data processing. OEC is also an exploration of edge computing in the context of space exploration. This section will elaborate on the basic concepts, typical systems, applications, and challenges associated with OEC.

6.4.1 Basic Concepts

Traditional satellites primarily function by relaying signals from ground stations, enabling long-distance data transmission, or by collecting observational data from space and sending it back to Earth. A commonly used technology in this context is the bent-pipe satellite transponder. The bent-pipe is a core component of communication satellites, primarily serving as an intermediary for data transmission, facilitating signal switching between uplink and downlink. In simpler terms, it forwards data from one side to the other, enabling communication between satellites and ground stations. The performance parameters of the bent-pipe transponder directly influence the overall performance of satellite communication systems. Therefore, traditional satellites have relatively weak data-processing capabilities and mainly function as switches or sensors.

Satellites are classified by their orbital heights into low Earth orbit (LEO), highly elliptical orbit (HEO), middle Earth orbit (MEO), and geostationary orbit (GEO). Table 6.3 from [41] provides the characteristics of satellites in different orbits, including altitude, period, and so on. In recent years, with the development of LEO technologies, especially with the introduction of StarLink, LEO satellites have gradually come into public focus. Currently, thousands of LEO satellites have been launched globally. These satellites are capable of providing around-the-clock network services to most regions worldwide, which has spurred interest in OEC.

In fact, as early as 2011, the concept of satellite cloud computing was proposed [16]. However, possibly due to the idea being too ahead of its time and the demand not yet being mature, it did not attract much attention. In recent years, with the improvement in satellite capabilities and the increasing amounts of data for sensing, relaying, and transmission, people have started to consider orbital edge computing. The goal is to use edge computing technology to reduce the amount of data communication between satellites and the ground, thereby enabling satellites to serve more users [8, 27].

The first concept and architecture of OEC were proposed by Denby and Lucia [8, 9] from Carnegie Mellon University in 2019. The work [9] demonstrated that OEC could improve the efficiency of remote sensing image processing by avoiding redundant data transmission between satellites and the ground. Today, OEC is considered a promising technology for various applications. For instance,

Table 6.3 Comparison of satellites in different orbit.

	LEO	HEO	MEO	GEO
Altitude	300–1500 km	600–40,000 km	8000–20,000 km	35,786 km
Period	1.4–2.5 h	12 h	6–12 h	24 h
Num. of satellites a constellation	24,000	4–8	8–16	3–4
Coverage	Global	High latitude areas	Global	Global (except polar regions)
Latency	5–35 ms	150–250 ms	50–100 ms	270 ms
Pass durations	About 10 min	4–8 h	1–2 h	All the time
Typical constellation	Iridium, Starlink, Kuiper, O3B	Molniya, Loopus, Archimedes	Odyssey	Inmarsat, MSAT, Mobilesat

Source: Data from Wu et al. [41].

remote sensing images captured by LEO satellites can be processed directly by OEC, thus reducing the amount of data that needs to be transmitted back to Earth.

6.4.2 Advanced Concepts and Architecture

In this section, we will further elaborate on the advantages of OEC. We will begin with a detailed analysis of the advantages in communication time using a case study. Next, we will introduce the first OEC architecture, followed by an introduction to two typical OEC models, in terms of end-edge collaboration and edge-edge collaboration.

6.4.2.1 Advantages of Orbital Edge Computing

If orbital edge computing is successfully implemented, we can envision satellites in space providing low-latency, globally covered services to ground users, thereby significantly enhancing the user experience across various services. Specifically:

- **Low latency**: Satellites, especially LEO satellites, can provide low-latency communication on a global scale. As shown in Table 6.3, the communication latency of LEO satellites typically ranges between 5 and 35 ms. In addition, Bhattacherjee et al. [2] evaluated the latency of two constellation networks, Starlink and Kuiper, and found that most latencies were around 4 ms, with some regions experiencing latencies between 8 and 16 ms. In contrast, for current terrestrial wired networks, even with CDN technology, network latency often reaches tens to hundreds of milliseconds. For some services without CDN acceleration, or in mobile scenarios, network latency can extend to several hundred milliseconds [2]. This higher latency is typically caused by numerous routers and switches forwarding data. However, with OEC, services are deployed directly on the satellites, allowing users to access them directly and avoiding multiple hops and the processing of data packets along the way.
- **Global coverage**: On the other hand, OEC brings the promise of computing to every corner of the Earth. Traditional cloud data centers are relatively sparse on the map, with some regions, such as South America and Africa, having little to no presence. In contrast, there are already thousands of LEO satellites in orbit (SpaceX's plan is the most ambitious, with the goal of launching 42,000 satellites). Although the computing power of these satellites is limited, they can at least provide basic network access and extend cloud functionality to OEC, bringing computational resources to regions across the globe. Therefore, OEC can offer ubiquitous "edge computing" without the many challenges of deploying ground infrastructure in various locations. In this case, some cloud computing providers have already started to follow this trend.

6.4.2.2 Typical Architectures in OEC

Based on the two advantages mentioned above, OEC shows significant potential. On the one hand, satellites can serve as edge nodes, processing sensing data to reduce the amount of data that needs to be transmitted to the ground, allowing the limited downlink bandwidth to support more satellites. On the other hand, satellites can form a cloud computing infrastructure or edge layer in space, providing computational resources and offering diverse services to users. In this subsection, we will introduce examples of typical system architecture designs based on these two main approaches. However, it is important to note that the progress of OEC is still relatively slow. This is primarily due to the challenges of constructing experimental environments, as there are currently not many satellite platforms available for independently developed applications.

Computing on Satellites In the traditional model, all satellite data is transmitted back to the ground for computation. However, in the end-edge collaboration model of OEC, part of the computation can be done on the satellite, which acts as an edge node. Through coordination between the edge nodes, the computational load on a single satellite can be significantly reduced, as well as the amount of data that needs to be transmitted back to Earth.

Recently, Denby and Lucia [9] designed an OEC architecture, which can be considered the first system architecture for OEC. This architecture primarily processes data on satellites, allowing only partial, relevant data to be transmitted to the ground, thereby alleviating the bottleneck of the downlink bandwidth. Additionally, the architecture enables the coordination of computational tasks between different satellites, leading to more optimized processing. To further enhance data processing, Denby et al. [10] also proposed an architecture called Kodan. As shown in Figure 6.6, Kodan's architecture is designed to work in two phases: prelaunch and postlaunch. Before launch, models can be customized based on the satellite's hardware, processing time, location, etc. Once deployed, the satellite uses these

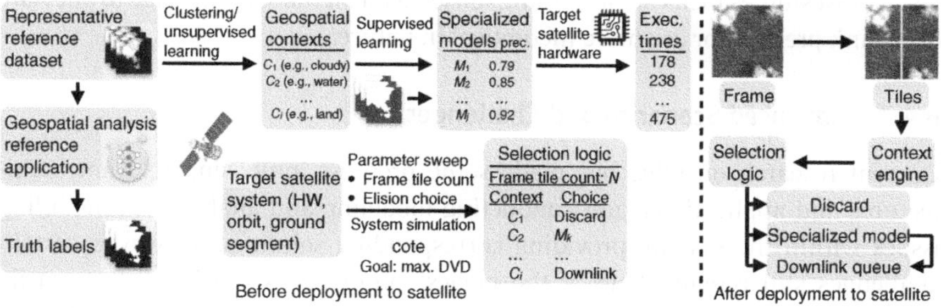

Figure 6.6 Kodan architecture design. Source: Denby et al. [10]/ACM, Inc.

models to extract high-value data from the sensed information, thereby maximizing the use of the limited downlink capacity.

Currently, there are many similar works. For example, Leyva-Mayorga et al. [19] applied OEC technology in real-time, ultra-high-resolution Earth observation applications to reduce the cost of image transmission. In their approach, they model the state of the downlink (such as bandwidth, connection time, etc.) along with the requirements for image quality and latency. Based on this model, they propose a scheduling algorithm to determine the image compression quality, ensuring that the transmission can meet the downlink conditions.

Collaborative Computing The aforementioned works primarily focus on improving data quality under limited downlink conditions by leveraging OEC. These works mainly consider computing on the orbit side but do not take into account the computational needs on the end side. In other words, the above works merely bring computation to the edge but do not fully implement the computational path envisioned in edge computing. Regarding edge collaboration, there are some existing studies, but most of them focus on resource scheduling optimization, with fewer works addressing system-level considerations.

Given the possibility that satellites may deploy edge services and provide computational resources to users, Zhang et al. [50] proposed the OEC Task Allocation (OEC-TA) algorithm. In their system model, users can upload tasks to a satellite via a ground station. The satellite then decomposes the tasks based on a greedy algorithm and schedules them for collaborative processing across the satellite constellation. The final results are sent back to the user, thereby enabling the deployment of edge services on orbiting satellites. Similarly, Liu et al. [22] proposed an advanced computation scheduling algorithm for OEC. In their system, satellites can offload computational tasks to other satellites or to the ground. Their approach mainly considers factors such as energy generated by solar exposure and the energy consumption of task computation, aiming to minimize the energy usage of satellites.

However, it is also clear that current research on OEC is still limited at the system level. Most studies focus on processing sensed data on the satellite edge to alleviate downlink pressure or are more theoretical in nature.

6.4.3 Advanced Scenarios and Challenges

Current research on OEC has made some progress, proposing relevant architectures and applications, and conducting modeling and analysis of scheduling issues within the system, providing corresponding solutions. However, certain challenges still remain. One of the biggest problems is the lack of an easily accessible platform for researchers to use. Nevertheless, this does not hinder our vision for the technology. If, in the future, satellite constellations are able to provide OEC services, as aforementioned, OEC could leverage its advantages in low latency and global coverage to greatly enhance our daily lives.

- **AR/VR**: One of the first applications to benefit from this would be augmented reality (AR)/VR-related applications. Currently, our VR/AR activities are mostly conducted on single devices, and due to the high computational demands of AR applications, rendering and data transmission have already introduced some latency overhead. As a result, it is still difficult for AR applications to support collaborative scenarios, such as remote gatherings. As previously mentioned, in the current Internet environment, most network transmissions have latencies ranging from tens to hundreds of milliseconds. This delay makes it challenging for multiple users to share interactions through cloud services without affecting the experience, particularly in gaming. By offloading some of the cloud service functions, such as interaction-sharing services, to satellites, it would be possible to transmit only a small amount of data while achieving low latency—such as the 5–35 ms shown in Table 6.3. This would greatly improve the user experience. Similarly, AR/VR would not be limited to gaming but could also extend to daily work activities, such as collaborative surgeries, design discussions, and more.
- **Space exploration**: In this section, we have discussed how OEC applied to Earth observation can bring numerous benefits, as demonstrated by the first OEC project. Going further, space exploration can also benefit from OEC. Currently, some space probes send their data back to Earth for analysis, and raw data inevitably consumes a significant amount of bandwidth. Additionally, the data transmission windows for some satellites are limited. By leveraging OEC, this data can be distributed to other nodes in space for relaying and processing, allowing for continuous data transmission around the clock while reducing the volume of data sent back, thereby addressing the downlink bandwidth limitations.

However, there are still numerous challenges facing OEC that require urgent research and resolution in order to fully harness the potential of edge computing in satellite-based computation.

- **System platforms and software interfaces**: The computing environment on satellites differs from that on the ground, and traditional cloud computing and edge computing orchestrators cannot be directly applied to satellites without encountering issues. Satellites themselves are essentially systems that integrate communication and computation, so a specialized operating system is needed to manage resources on such hardware systems. From a research perspective, there is currently no universal platform or easy-to-use simulator available for researchers, making it difficult for most to conduct related studies.
- **Resource management**: While LEO satellites can provide low-latency communication, the time they can serve a specific region (or user) is extremely limited, often just a matter of minutes. Although satellite constellations can provide alternating services, this places stringent demands on scheduling algorithms. On the one hand, satellite positioning and user relationships need to be modeled, and on the other hand, modeling the airborne satellite network is necessary

to optimize scheduling. Furthermore, OEC must offer seamless data link migration without any user-perceived interruptions, or else the user experience will be severely impacted.

6.5 Summary and Practice

6.5.1 Summary

Although several new computing paradigms have been proposed in recent years, such as sky computing, computing power network, and meta computing, they can essentially be viewed as enhancements to edge computing in certain aspects. In other words, they can all be integrated with edge computing to further enhance its capabilities. For example, sky computing emphasizes the collaboration of heterogeneous clouds, computing power network focuses on the synergy between edge and multi-cloud environments, while meta computing primarily addresses the security issues in cross-domain collaboration within edge computing.

Next, we introduced the application of edge computing in several emerging technologies. In the field of artificial intelligence, edge computing can enhance the intelligence of devices on the edge by training, inferring, and caching large language models at the network edge. In the 6G domain, edge computing serves as the computational foundation of the 6G network architecture, providing a base platform for AI technologies to achieve intelligent signal control and network slicing. At the same time, 6G represents a significant application scenario for edge computing. In the field of orbital computing, some satellites, particularly low-orbit satellites, are equipped with limited computational capabilities, giving rise to the emerging field of orbital edge computing. In this area, edge computing is still in its infancy; the primary focus is on processing data at the edge on satellites to reduce data transmission costs and to empower space exploration.

Finally, in the emerging technologies mentioned earlier, we also discussed several research directions that are worth further exploration, with the hope of encouraging more researchers to engage in systematic studies in these areas.

6.5.2 Practice Questions

1) What is the relationship between edge computing and a new distributed computing paradigm, using one as an example?
2) How does edge computing enhance artificial intelligence, particularly large language model technology?
3) What are the main characteristics of 6G, and how do 6G and edge computing technologies influence each other?
4) Discuss the challenges and opportunities of integrating orbital computing technology with edge computing.

6.5.3 Course Projects

1) Deploy large language models, such as Edge-LLM (https://github.com/GATECH-EIC/Edge-LLM), LlamaEdge (https://github.com/LlamaEdge/LlamaEdge), on heterogeneous hardware and explore the segmentation of specific models at different layers in an edge computing environment, observing the changes in transmission costs and computational costs.
2) Inspired by the concepts discussed in the paper "The Internet of Things in the Era of Generative AI: Vision and Challenges," Wang et al. [40] select a specific IoT application (e.g., smart home, healthcare monitoring, and industrial IoT) and explore how generative AI can be applied to enhance its functionality.
3) By leveraging open-source projects, construct scenarios for cross-cloud and cross-domain edge collaboration. Consider using communication networks from different telecom operators to simulate cross-domain edge collaboration, and try to understand the distinctions between sky computing, computing power network, and meta computing in comparison to edge computing. Recommended open-source projects include: OpenFaaS, KubeEdge, and Kubernetes.
4) Utilize the open-source project Cote (https://github.com/CMUAbstract/cote) to explore the principles and applications of orbital edge computing.

Chapter 6 Suggested Papers

1 Xiuzhen Cheng et al. "Meta computing". In: *IEEE Network* 38. 2 (2024), pp. 225–231. ISSN: 1558-156X. DOI: 10.1109/MNET003.2300092.

2 Bradley Denby and Brandon Lucia. "Orbital edge computing: Nanosatellite constellations as a new class of computer system". In: *Proceedings of the 25th International Conference on Architectural Support for Programming Languages and Operating Systems*. ASPLOS '20. New York, NY, USA: Association for Computing Machinery, Mar. 2020, pp. 939–954. ISBN: 978-1-4503-7102-5. DOI: 10.1145/3373376.3378473.

3 Khaled B Letaief et al. "Edge artificial intelligence for 6G: Vision, enabling technologies, and applications". In: *IEEE Journal on Selected Areas in Communications* 40. 1 (2022), pp. 5–36. ISSN: 1558-0008. DOI: 10.1109/JSAC.2021.3126076.

4 Guanqiao Qu et al. *Mobile Edge Intelligence for Large Language Models: A Contemporary Survey*. arXiv:2407.18921 [cs]. July 2024. DOI: 10.48550/arXiv.2407.18921. url: http://arxiv.org/abs/2407.18921.

5 Ion Stoica and Scott Shenker. "From cloud computing to sky computing". In: *HotOS '21: Workshop on Hot Topics in Operating Systems, Ann Arbor, Michigan,*

USA, June, 1–3, 2021. Ed. by Sebastian Angel, Baris Kasikci, and Eddie Kohler. ACM, 2021, pp. 26–32. DOI: 10.1145/3458336.3465301.

References

1 Tiemo Bang et al. “SkyPIE: A fast & accurate oracle for object placement”. In: *Proceedings of the ACM on Management of Data* 2. 1 (2024), pp. 55:1–55:27. DOI: 10.1145/3639310.

2 Debopam Bhattacherjee et al. “In-orbit computing: An outlandish thought experiment?” In: *Proceedings of the 19th ACM Workshop on Hot Topics in Networks*. HotNets ’20. New York, NY, USA: Association for Computing Machinery, Nov. 2020, pp. 197–204. ISBN: 978-1-4503-8145-1. DOI: 10.1145/3422604.3425937.

3 Qingqing Cao et al. *BTR: Binary Token Representations for Efficient Retrieval Augmented Language Models*. May 2024. DOI: 10.48550/arXiv.2310.01329. arXiv: 2310.01329 [cs].

4 Yanxi Chen et al. *EE-LLM: Large-Scale Training and Inference of Early-Exit Large Language Models with 3D Parallelism*. June 2024. DOI: 10.48550/arXiv.2312.04916. arXiv: 2312.04916 [cs].

5 Xiuzhen Cheng et al. “Meta computing”. In: *IEEE Network* 38. 2 (2024), pp. 225–231. ISSN: 1558-156X. DOI: 10.1109/MNET003.2300092.

6 Rong Cong et al. “EdgeGO: A mobile resource-sharing framework for 6G edge computing in massive IoT systems”. In: *IEEE Internet of Things Journal* 9. 16 (2022), pp. 14521–14529. ISSN: 2327-4662. DOI: 10.1109/JIOT.2021.3065357. URL: https://ieeexplore.ieee.org/abstract/document/9375469.

7 Can Cui et al. “Receive, reason, and react: Drive as you say, with large language models in autonomous vehicles”. In: *IEEE Intelligent Transportation Systems Magazine* 16. 4 (2024), pp. 81–94. ISSN: 1941-1197. DOI: 10.1109/MITS.2024.3381793. URL: https://ieeexplore.ieee.org/abstract/document/10491134.

8 Bradley Denby and Brandon Lucia. “Orbital edge computing: Machine inference in space”. In: *IEEE Computer Architecture Letters* 18. 1 (2019), pp. 59–62. ISSN: 1556-6064. DOI: 10.1109/LCA.2019.2907539.

9 Bradley Denby and Brandon Lucia. “Orbital edge computing: Nanosatellite constellations as a new class of computer system”. In: *Proceedings of the 25th International Conference on Architectural Support for Programming Languages and Operating Systems*. ASPLOS ’20. New York, NY, USA: Association for Computing Machinery, Mar. 2020, pp. 939–954. ISBN: 978-1-4503-7102-5. DOI: 10.1145/3373376.3378473.

10 Bradley Denby et al. “Kodan: Addressing the computational bottleneck in space”. In: *Proceedings of the 28th ACM International Conference on*

Architectural Support for Programming Languages and Operating Systems, Volume 3. ASPLOS 2023. New York, NY, USA: Association for Computing Machinery, Mar. 2023, pp. 392–403. ISBN: 978-1-4503-9918-0. DOI: 10.1145/3582016.3582043.

11 Nuria González-Prelcic et al. "The integrated sensing and communication revolution for 6G: Vision, techniques, and applications". In: *Proceedings of the IEEE* 112. 7 (2024), pp. 676–723. ISSN: 1558-2256. DOI: 10.1109/JPROC.2024.3397609.

12 Ying He et al. "Large language models (LLMs) inference offloading and resource allocation in cloud-edge computing: An active inference approach". In: *IEEE Transactions on Mobile Computing* 23. 12 (2024), pp. 11253–11264. ISSN: 1558-0660. DOI: 10.1109/TMC.2024.3415661. URL: https://ieeexplore.ieee.org/abstract/document/10591707.

13 Hui Huang, Qiang Ye, and Yitong Zhou. "6G-empowered offloading for real-time applications in multi-access edge computing". In: *IEEE Transactions on Network Science and Engineering* 10. 3 (2023), pp. 1311–1325. ISSN: 2327-4697. DOI: 10.1109/TNSE.2022.3188921. URL: https://ieeexplore.ieee.org/abstract/document/9817805.

14 Paras Jain et al. "Skyplane: Optimizing transfer cost and throughput using cloud-aware overlays". In: *20th USENIX Symposium on Networked Systems Design and Implementation (NSDI 23)*. 2023, pp. 1375–1389. ISBN: 978-1-939133-33-5.

15 Frank Joublin et al. "CoPAL: Corrective planning of robot actions with large language models". In: *2024 IEEE International Conference on Robotics and Automation (ICRA)*. May 2024, pp. 8664–8670. DOI: 10.1109/ICRA57147.2024.10610434. URL: https://ieeexplore.ieee.org/abstract/document/10610434.

16 Kamen Kanev and Nikolay Mirenkov. "Satellite cloud computing". In: *2011 IEEE Workshops of International Conference on Advanced Information Networking and Applications*. Mar. 2011, pp. 147–152. DOI: 10.1109/ WAINA.2011.61.

17 Yiping Kang et al. "Neurosurgeon: Collaborative intelligence between the cloud and mobile edge". In: *ACM SIGARCH Computer Architecture News* 45. 1 (2017), pp. 615–629. ISSN: 0163-5964. DOI: 10.1145/3093337.3037698.

18 Khaled B Letaief et al. "The roadmap to 6G: AI empowered wireless networks". In: *IEEE Communications Magazine* 57. 8 (2019), pp. 84–90. ISSN: 1558-1896. DOI: 10.1109/MCOM.2019.1900271.

19 Israel Leyva-Mayorga et al. "Satellite edge computing for real-time and very-high resolution earth observation". In: *IEEE Transactions on Communications* 71. 10 (2023), pp. 6180–6194. ISSN: 1558-0857. DOI: 10.1109/TCOMM.2023.3296584.

20 Zheng Lin et al. "Efficient parallel split learning over resource-constrained wireless edge networks". In: *IEEE Transactions on Mobile Computing* (2024), pp. 1–16. ISSN: 1558-0660. DOI: 10.1109/TMC.2024.3359040. URL: https://ieeexplore.ieee.org/abstract/document/10415235.

21 Fan Liu et al. "Integrated sensing and communications: Toward dual-functional wireless networks for 6G and beyond". In: *IEEE Journal on Selected Areas in Communications* 40. 6 (2022), pp. 1728–1767. ISSN: 1558-0008. DOI: 10.1109/JSAC.2022.3156632. URL: https://ieeexplore.ieee.org/abstract/document/9737357.

22 Weisen Liu et al. "In-orbit processing or not? Sunlight-aware task scheduling for energy-efficient space edge computing networks". In: *IEEE INFOCOM 2024 - IEEE Conference on Computer Communications*. May 2024, pp. 881–890. DOI: 10.1109/INFOCOM52122.2024.10621268.

23 Ruilong Ma et al. "Poster: PipeLLM: Pipeline LLM inference on heterogeneous devices with sequence slicing". In: *Proceedings of the ACM SIGCOMM 2023 Conference*. ACM SIGCOMM '23. New York, NY, USA: Association for Computing Machinery, Sept. 2023, pp. 1126–1128. DOI: 10.1145/3603269.3610856.

24 Mithun Mukherjee et al. *The Interplay of Reconfigurable Intelligent Surfaces and Mobile Edge Computing in Future Wireless Networks: A Win-Win Strategy to 6G*. arXiv:2106.11784 [cs, math]. May 2021. DOI: 10.48550/arXiv.2106.11784. URL: http://arxiv.org/abs/2106.11784.

25 Deepak Narayanan et al. "Efficient large-scale language model training on GPU clusters using megatron-LM". In: *Proceedings of the International Conference for High Performance Computing, Networking, Storage and Analysis*. SC '21. New York, NY, USA: Association for Computing Machinery, Nov. 2021, pp. 1–15. ISBN: 978-1-4503-8442-1. DOI: 10.1145/3458817.3476209. URL: https://dl.acm.org/doi/10.1145/3458817.3476209.

26 Shoki Ohta and Takayuki Nishio. *Λ-Split: A Privacy-Preserving Split Computing Framework for Cloud-Powered Generative AI*. Oct. 2023. DOI: 10.48550/arXiv.2310.14651. arXiv: 2310.14651 [cs].

27 Tobias Pfandzelter. "Serverless abstractions for edge computing in large low-earth orbit satellite networks". In: *Proceedings of the 24th International Middleware Conference: Demos, Posters and Doctoral Symposium*. Middleware '23. New York, NY, USA: Association for Computing Machinery, Dec. 2023, pp. 3–6. DOI: 10.1145/3626564.3629088.

28 Guanqiao Qu et al. *TrimCaching: Parameter-sharing Edge Caching for AI Model Downloading*. arXiv:2404.14204 [cs]. May 2024. DOI: 10.48550/arXiv.2404.14204. URL: http://arxiv.org/abs/2404.14204.

29 Sandeep Singh Sengar et al. "*Generative Artificial Intelligence: A Systematic Review and Applications*". In: *arXiv preprint arXiv:2405.11029* (2024).

30 Yifei Shen et al. "Large language models empowered autonomous edge AI for connected intelligence". In: *IEEE Communications Magazine* (2024), pp. 1–7. ISSN: 1558-1896. DOI: 10.1109/MCOM.001.2300550. URL: https://ieeexplore.ieee.org/abstract/document/10384606.

31 Ion Stoica and Scott Shenker. "From cloud computing to sky computing". In: *HotOS '21: Workshop on Hot Topics in Operating Systems, Ann Arbor, Michigan, USA, June, 1–3, 2021*. Ed. by Sebastian Angel, Baris Kasikci, and Eddie Kohler. ACM, 2021, pp. 26–32. DOI: 10.1145/3458336.3465301.

32 Wankai Tang et al. "Wireless communications with reconfigurable intelligent surface: Path loss modeling and experimental measurement". In: *IEEE Transactions on Wireless Communications* 20. 1 (2021), pp. 421–439. ISSN: 1558-2248. DOI: 10.1109/TWC. 2020.3024887. URL: https://ieeexplore.ieee.org/abstract/document/9206044.

33 Xiongyan Tang et al. "Computing power network: The architecture of convergence of computing and networking towards 6G requirement". In: *China Communications* 18. 2 (2021), pp. 175–185. ISSN: 1673-5447. DOI: 10.23919/JCC.2021.02.011.

34 Harish Viswanathan and Preben E Mogensen. "Communications in the 6G era". In: *IEEE Access* 8 (2020), pp. 57063–57074. ISSN: 2169-3536. DOI: 10.1109/ACCESS.2020.2981745.

35 Harish Viswanathan and Preben E Mogensen. "Communications in the 6G era". In: *IEEE Access* 8 (2020), pp. 57063–57074. ISSN: 2169-3536. DOI: 10.1109/ACCESS.2020.2981745. URL: https://ieeexplore.ieee.org/abstract/document/9040431.

36 Cheng-Xiang Wang et al. "On the road to 6G: Visions, requirements, key technologies, and testbeds". In: *IEEE Communications Surveys & Tutorials* 25. 2 (2023), pp. 905–974. ISSN: 1553-877X. DOI: 10.1109/COMST.2023. 3249835.

37 Yiding Wang et al. "Tabi: An efficient multi-level inference system for large language models". In: *Proceedings of the 18th European Conference on Computer Systems*. EuroSys '23. New York, NY, USA: Association for Computing Machinery, May 2023, pp. 233–248. ISBN: 978-1-4503-9487-1. DOI: 10.1145/3552326.3587438.

38 Guanqun Wang et al. "Cloud-device collaborative learning for multimodal large language models". In: *Proceedings of the IEEE/CVF Conference on Computer Vision and Pattern* Recognition. 2024, pp. 12646–12655. URL: https://openaccess.thecvf.com/content/CVPR2024/html/Wang_Cloud-Device: Collaborative_Learning_for_Multimodal_Large_Language_Models_CVPR_2024_paper.html.

39 Jiaqi Wang et al. *Large Language Models for Robotics: Opportunities, Challenges, and Perspectives*. arXiv:2401.04334 [cs]. Jan. 2024. DOI: 10.48550/arXiv.2401.04334. URL: http://arxiv.org/abs/2401.04334.

40 Xin Wang et al. "The Internet of Things in the era of generative AI: Vision and challenges". In: *IEEE Internet Computing* (2024). DOI: 10.1109/MIC.2024.3443169.

41 Changhao Wu et al. *A Comprehensive Survey on Orbital Edge Computing: Systems, Applications, and Algorithms*. June 2023. DOI: 10.48550/arXiv.2306.00275. arXiv: 2306.00275 [cs].

42 Daliang Xu et al. *LLMCad: Fast and Scalable On-device Large Language Model Inference*. Sept. 2023. DOI: 10.48550/arXiv.2309.04255. URL: http://arxiv.org/abs/2309.04255.

43 Minrui Xu et al. "Unleashing the power of edge-cloud generative AI in mobile networks: A survey of AIGC services". In: *IEEE Communications Surveys & Tutorials* 26. 2 (2024), pp. 1127–1170. ISSN: 1553-877X. DOI: 10.1109/COMST.2024.3353265. URL: https://ieeexplore.ieee.org/abstract/document/10398474.

44 Zhenhua Xu et al. "DriveGPT4: Interpretable end-to-end autonomous driving via large language model". In: *IEEE Robotics and Automation Letters* 9. 10 (2024), pp. 8186–8193. ISSN: 2377-3766. DOI: 10.1109/LRA.2024.3440097. URL: https://ieeexplore.ieee.org/abstract/document/10629039.

45 Zongheng Yang et al. "SkyPilot: An intercloud broker for sky computing". In: *20th USENIX Symposium on Networked Systems Design and Implementation, NSDI 2023, Boston, MA, April 17–19, 2023*. Ed. by Mahesh Balakrishnan and Manya Ghobadi. USENIX Association, 2023, pp. 437–455.

46 Feng Ye et al. "Intelligent hierarchical NOMA-based network slicing in cell-free RAN for 6G systems". In: *IEEE Transactions on Wireless Communications* 23. 5 (2023), pp. 4724–4737. https://ieeexplore.ieee.org/abstract/document/10278091/.

47 Sun Yukun et al. "Computing power network: A survey". In: *China Communications* 21. 9 (2024), pp. 109–145. ISSN: 1673-5447. DOI: 10.23919/JCC.ja.2021-0776.

48 Qingyang Zhang et al. "Demo abstract: EVAPS: Edge video analysis for public safety". In: *2016 IEEE/ACM Symposium on Edge Computing (SEC)*. IEEE, 2016, pp. 121–122.

49 Quan Zhang et al. "Firework: Data processing and sharing for hybrid cloud-edge analytics". In: *IEEE Transactions on Parallel and Distributed Systems* 29. 9 (2018), pp. 2004–2017. ISSN: 1558-2183. DOI: 10.1109/TPDS.2018.2812177.

50 Yuru Zhang et al. "Aerial edge computing on orbit: A task offloading and allocation scheme". In: *IEEE Transactions on Network Science and Engineering* 10. 1 (Jan. 2023), pp. 275–285. ISSN: 2327-4697. DOI: 10.1109/TNSE.2022.3207214.

51 Wayne Xin Zhao et al. *A Survey of Large Language Models*. arXiv:2303.18223 [cs]. Nov. 2023. DOI: 10.48550/arXiv.2303.18223. URL: http://arxiv.org/abs/2303.18223.

7

Case Studies and Practical Applications*

Edge computing represents a significant paradigm shift in the way data is processed, analyzed, and acted upon, bringing computation and data storage closer to the source of data generation. This chapter delves into an in-depth exploration of edge computing systems through the lens of real-world case studies across six pivotal sectors: manufacturing, Internet of Things (IoT) retail, healthcare, telecommunications, autonomous vehicles, and smart cities. By examining the implementation and outcomes in these diverse fields, we aim to illustrate the profound impact of edge computing on operational efficiency, decision-making, and innovation.

In the manufacturing sector, edge computing is revolutionizing production lines, enabling predictive maintenance, real-time quality control, and efficient resource management. By processing data at the edge, manufacturers can reduce downtime, enhance productivity, and ensure higher standards of product quality. The deployment of edge computing in smart factories exemplifies how leveraging real-time data can lead to substantial cost savings and improved operational resilience [32].

IoT significantly benefits from edge computing by enhancing real-time data processing, reducing latency, and improving overall system efficiency. In IoT ecosystems, vast amounts of data are generated by numerous connected devices, often requiring immediate analysis and response. Edge computing addresses this need by processing data closer to the source, minimizing the delay caused by data transmission to centralized cloud servers. This proximity to data sources not only reduces network congestion but also ensures faster decision-making and more timely actions, which are critical in applications such as autonomous vehicles, industrial automation, and healthcare monitoring. Additionally, edge

*This chapter is contributed by Yuankai He.

Edge Computing: Systems and Applications, First Edition. Lanyu Xu and Weisong Shi.

Companion Website: www.wiley.com/go/xu/computing

computing enhances data security and privacy by limiting the exposure of sensitive information to potential vulnerabilities associated with cloud storage and transmission. By decentralizing data processing and bringing computational power closer to IoT devices, edge computing enables more robust, scalable, and responsive IoT solutions [5].

Retail, on the other hand, benefits from edge computing by enhancing customer experiences through personalized services and smarter inventory management. By analyzing data locally, retailers can offer customized promotions, optimize supply chains, and respond swiftly to market trends. The primary problem is the increasing data volume from IoT devices, which creates challenges in managing, analyzing, and utilizing data in real-time, especially for brick-and-mortar stores competing with online retailers. Key challenges include ensuring low latency, managing computational load, and integrating various IoT technologies effectively [8].

Healthcare stands to gain immensely from edge computing, particularly in the realms of patient monitoring, diagnostics, and treatment. Edge computing allows for real-time processing of medical data from wearable devices, facilitating timely interventions and personalized care. In this chapter, we will examine case studies demonstrating how edge computing supports advanced healthcare applications, from remote patient monitoring to enhanced medical imaging, ultimately leading to better patient outcomes and more efficient healthcare delivery systems [15, 38].

The telecommunications industry is at the forefront of edge computing adoption, driven by the need to support the ever-growing demand for bandwidth and low-latency services. Edge computing plays a crucial role in optimizing network performance, enabling the deployment of 5G networks, and supporting emerging applications such as augmented reality (AR) and virtual reality (VR). Through detailed case studies, we will uncover how telecommunications companies leverage edge computing to enhance service delivery, improve network efficiency, and drive innovation in communication technologies [51]. A summary of potential edge computing applications can be summarized in Figure 7.1.

Autonomous vehicles represent a cutting-edge application of edge computing, where the need for real-time data processing is paramount. Edge computing enables autonomous vehicles to make real-time decisions based on local data, ensuring safety and reliability. Challenges include ensuring low-latency data processing, maintaining energy efficiency, and securing the system against potential attacks across various layers. Edge computing systems can be used to handle the intensive computational tasks locally, reducing the dependency on centralized cloud infrastructure. Additionally, vehicle-to-everything (V2X) communication is highlighted as a critical technology for providing redundancy and alleviating computational load, thereby enhancing the overall reliability and safety of autonomous driving systems [25, 28].

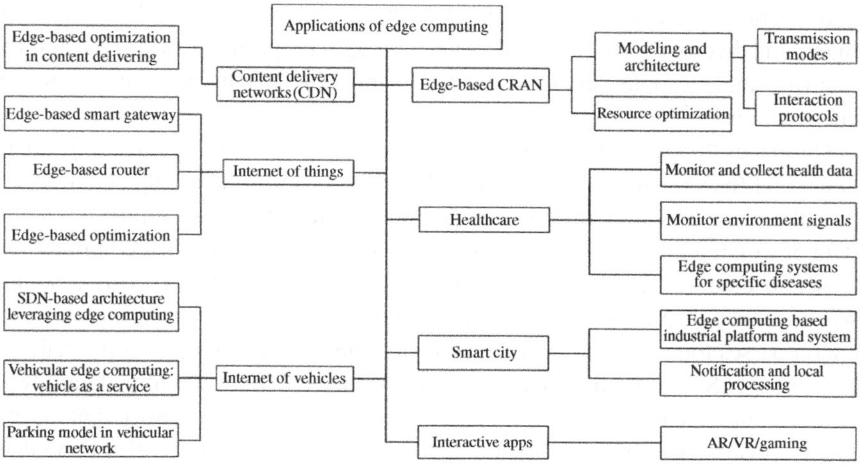

Figure 7.1 A taxonomy of edge computing applications. Source: Zhao et al. [51]/IEEE.

Lastly, smart cities epitomize the transformative potential of edge computing on urban living. By deploying edge computing systems, cities can manage resources more effectively, enhance public safety, and improve the quality of life for their inhabitants. From intelligent traffic management to smart energy grids, this chapter will explore how edge computing facilitates the creation of more sustainable and livable urban environments [3, 14, 20, 34, 40].

Through these comprehensive case studies, this chapter aims to provide a thorough understanding of how edge computing is being implemented across various industries. We will analyze the challenges faced, solutions devised, and the tangible benefits realized by each sector. By bridging the gap between theoretical concepts and practical applications, this chapter serves as a crucial resource for understanding the role of edge computing in driving forward technological advancements and shaping the future of these critical domains [16].

7.1 Manufacturing

Edge computing significantly enhances the manufacturing sector by improving efficiency, reducing latency, and enabling real-time data processing. By processing data closer to the source, manufacturers can achieve rapid and autonomous decision-making, essential for intelligent manufacturing systems. This approach facilitates predictive maintenance, timely detection and response to production anomalies, and real-time quality control, leading to reduced downtime and increased productivity. Additionally, edge computing optimizes resource management and bandwidth usage, ensuring higher standards of product quality and operational resilience. Despite challenges like middleware flexibility and managing diverse communication protocols, edge computing provides the agility, security, and responsiveness needed for modern, IoT-based manufacturing environments.

The paper "Edge Computing in IoT-Based Manufacturing" by Baotong Chen et al. [11] (as shown in Figure 7.2) explores the pivotal role of edge computing in enhancing operational efficiency, reducing latency, and enabling real-time data processing in the manufacturing sector. Edge computing shifts computation closer to data sources, which is critical for intelligent manufacturing systems requiring rapid and autonomous decision-making. The proposed architecture encompasses four domains: devices (sensors, robots), network (software-defined networking [SDN] and time-sensitive networking [TSN] for real-time data flow), data (cleaning, feature extraction), and applications (intelligent process management). A case study on active maintenance in a smart factory revealed that edge computing significantly improved efficiency, agility, and reduced network load by 60%, showcasing its potential for business agility and bandwidth optimization. Despite its benefits, challenges such as the need for flexible middleware, managing diverse communication protocols, and ensuring real-time processing with security remain. Overall, edge computing is essential for advancing IoT-based manufacturing, supporting the development of responsive and resilient industrial systems by providing enhanced agility, security, and real-time processing capabilities.

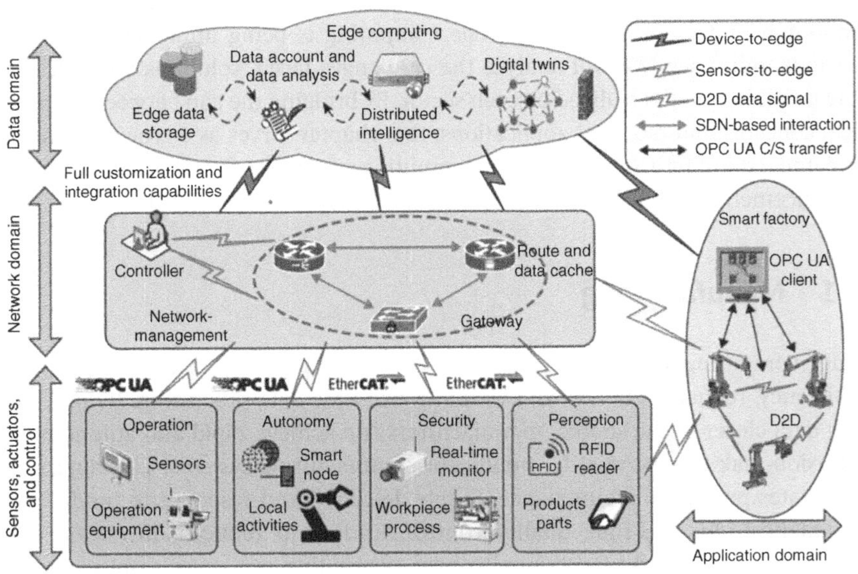

Figure 7.2 Edge computing in manufacturing. Source: Chen et al. [11]/IEEE.

The paper titled "Edge Computing Enabled Production Anomalies Detection and Energy-Efficient Production Decision Approach for Discrete Manufacturing Workshops" [49] addresses the complexities and dynamics of modern manufacturing processes, which frequently experience production anomalies such as spindle failure and cutting tool wear. These anomalies significantly impact manufacturing quality and productivity. The main challenge lies in the timely detection and response to these anomalies amidst the rapid development and data proliferation driven by IoT technologies. To tackle this, the paper proposes an innovative approach leveraging edge computing. The solution introduces a three-layer architecture for anomaly detection and energy-efficient production decisions, using an energy consumption data preprocessing algorithm and a production anomaly analysis model based on a long short-term memory (LSTM) network. This framework ensures real-time data processing, reduces latency, and supports energy-efficient decision-making when anomalies are detected. The proposed method demonstrates a high detection accuracy with an anomaly detection error of only 3.5%, proving its effectiveness in enhancing production process monitoring and energy conservation in a discrete manufacturing workshop.

The paper titled "Edge Computing in Smart Production" by Jumyung Um et al. [43] explores the implementation of edge computing within Cyber-Physical Production Systems to enhance flexibility and efficiency in smart manufacturing. The main problem addressed is the synchronization between digital models and physical objects, and the application of decision-making within these models, which is often hindered by unstable cloud connections and high latency. The challenges lie in managing the vast amounts of data generated by manufacturing processes, ensuring real-time response, and maintaining system reliability despite limited computing resources at the edge. The proposed solution involves an edge computing architecture that acts as an intermediary between machines, offering local cloud services with fast response times and preprocessing capabilities. This architecture supports real-time data processing and human–machine interaction, demonstrated through the preprocessing of data from AR devices to facilitate real-time communication with the cyber-model. The edge platform effectively manages computing resources and prioritizes processes, enabling dynamic updates to production lines and improving the overall efficiency and responsiveness of smart production environments.

In 2023, Yu et al.'s paper "Edge Computing-Assisted IoT Framework with an Autoencoder for Fault Detection in Manufacturing Predictive Maintenance" [48] addresses the urgent need for real-time, intelligent predictive maintenance in industrial manufacturing, which requires low latency responses to alarms. Traditional cloud-based IoT frameworks suffer from high latency and network congestion issues. The proposed solution integrates edge computing to decentralize data processing, reduce network pressure, and protect user privacy while optimizing cloud costs and resources. The framework introduces an autoencoder-based deep learning method for more accurate fault detection, implemented within a three-layer architecture consisting of edge, cloud, and application layers. This architecture facilitates real-time data ingestion, preprocessing, and analysis, enabling timely responses to maintenance needs. A distributed stacked sparse autoencoder is employed to handle the complex, nonlinear relationships between sensors, providing robust fault detection in a real-time, distributed manner. This approach significantly reduces system response time and enhances the performance and efficiency of the predictive maintenance ecosystem, making it a practical and scalable solution for smart manufacturing.

7.2 Telecommunications

Edge computing greatly benefits the telecommunications industry by improving performance metrics and reducing costs. It enhances throughput, reduces latency, and improves video delay by distributing applications and content closer to end-users, thus minimizing network congestion and enhancing Quality of Experience (QoE). This approach also reduces the total cost of ownership (TCO) by offloading peak traffic from the core network to edge nodes. Additionally, integrating edge computing with ultra-reliable low-latency communication (URLLC) addresses the high latency and reliability challenges of centralized cloud computing. It supports mission-critical applications like VR, V2X, and edge artificial intelligence (AI) by bringing computational resources closer to network nodes. This ensures timely and reliable data processing, leveraging technologies like high-capacity millimeter-wave links, proximity-based computing, and edge machine learning to meet stringent latency and reliability requirements. Overall, edge computing is crucial for advancing telecommunications infrastructure, optimizing network performance, and enabling new services.

The paper "Edge Cloud Computing in Telecommunications: Case Studies on Performance Improvement and TCO Saving" [12] (as shown in Figure 7.3) discusses the implementation and benefits of Edge Cloud Computing (ECC) in telecommunications networks. ECC is becoming crucial as new services like 4K/8K video streaming, 360-degree augmented/VR, and autonomous driving demand stringent key performance indicators (KPIs) and lower TCO. The paper provides several case studies demonstrating how ECC can improve KPIs such as throughput, latency, and video delay, by distributing applications and content closer to end-users, thus reducing network congestion and enhancing QoE.

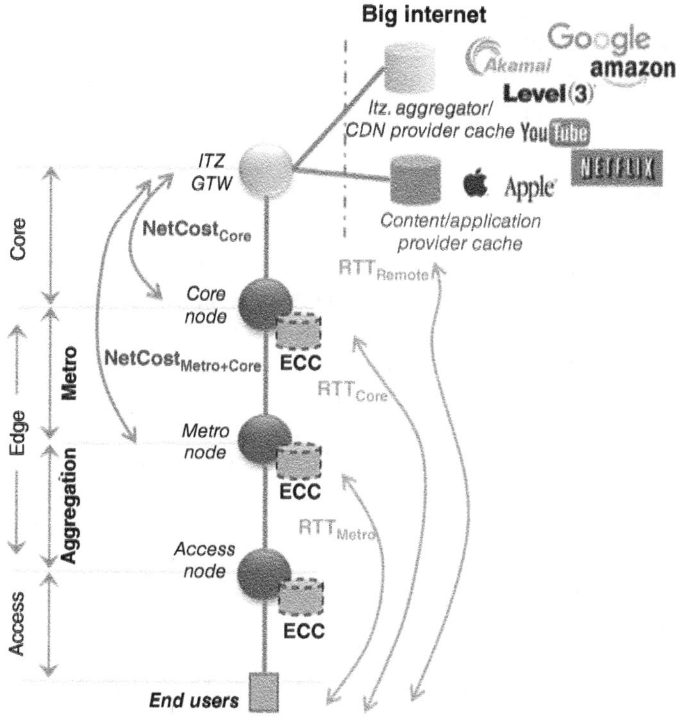

Figure 7.3 Edge computing in telecommunications. Source: Ciccarella et al. [12]/IEEE.

Additionally, ECC helps in reducing TCO by offloading peak traffic from the core network to edge nodes, leading to cost savings. The paper concludes that the strategic deployment of ECC can effectively address both performance and cost issues in traditional and emerging broadband networks, marking a significant step forward for telecommunications infrastructure.

The paper "Wireless Edge Computing With Latency and Reliability Guarantees" [19] investigates the feasibility and potential of integrating edge computing with ultra-reliable low-latency communication (URLLC) for mission-critical applications such as VR, V2X, and edge artificial intelligence (AI). The main problem addressed is the high latency and reliability challenges in centralized cloud computing architectures, which are inadequate for latency-sensitive applications. The proposed solution is a distributed edge computing architecture that brings computational and storage resources closer to end network nodes, significantly reducing latency and enhancing reliability. The paper explores several enablers for achieving low latency and high reliability, including high-capacity millimeter-wave links, proximity-based computing, edge machine learning, proactive computing, and parallel and coded computing. Through various use cases, including VR and vehicular edge computing, the paper demonstrates how edge computing can meet the stringent requirements of URLLC, ensuring timely and reliable data processing and decision-making at the network edge.

7.3 Healthcare

Edge computing significantly enhances healthcare systems (as shown in Figure 7.4) in smart cities by enabling real-time data processing, reducing latency, and improving responsiveness. By processing data closer to the source, edge computing addresses inefficiencies in traditional cloud-based systems, which struggle with high latency and bandwidth usage due to the vast amounts of data generated by IoT devices. This approach ensures timely healthcare analytics and decision-making, essential for applications like remote patient monitoring, predictive maintenance, and telemedicine. Edge computing also enhances data security and privacy by minimizing the exposure of sensitive health information to potential vulnerabilities associated with cloud storage and transmission. In elderly care, edge computing supports efficient, real-time health monitoring, empowering seniors to manage their health independently and reducing the burden on healthcare systems. By integrating edge computing with IoMT, healthcare services become more responsive and efficient, ensuring better patient outcomes and optimized resource usage. Edge computing thus holds significant potential to transform healthcare delivery in smart cities by improving the efficiency, security, and real-time capabilities of healthcare applications.

The paper titled "The Role of Edge Computing in Real-Time Analytics for Smart City Healthcare Applications" by Balaram Yadav Kasula [24] delves into the critical function of edge computing in enhancing real-time analytics in smart city healthcare contexts. The primary problem addressed is the inefficiency of traditional cloud-based systems in managing the vast amounts of data generated by IoT devices in urban health systems, leading to high latency and bandwidth usage issues. The challenges include the integration of edge devices with existing healthcare infrastructures, ensuring data security, and improving the responsiveness of healthcare analytics. The proposed solution involves leveraging edge computing to process data closer to the source, thereby reducing latency and bandwidth usage. The paper provides insights into the technical aspects of edge computing, emphasizing its capacity to enhance the efficiency and responsiveness of healthcare decision-making. Case studies demonstrate the successful implementation of edge computing, highlighting its potential to transform urban healthcare delivery by enabling quicker decision-making and improved patient outcomes. This research contributes valuable knowledge to the optimization of smart city healthcare systems through strategic edge computing deployment.

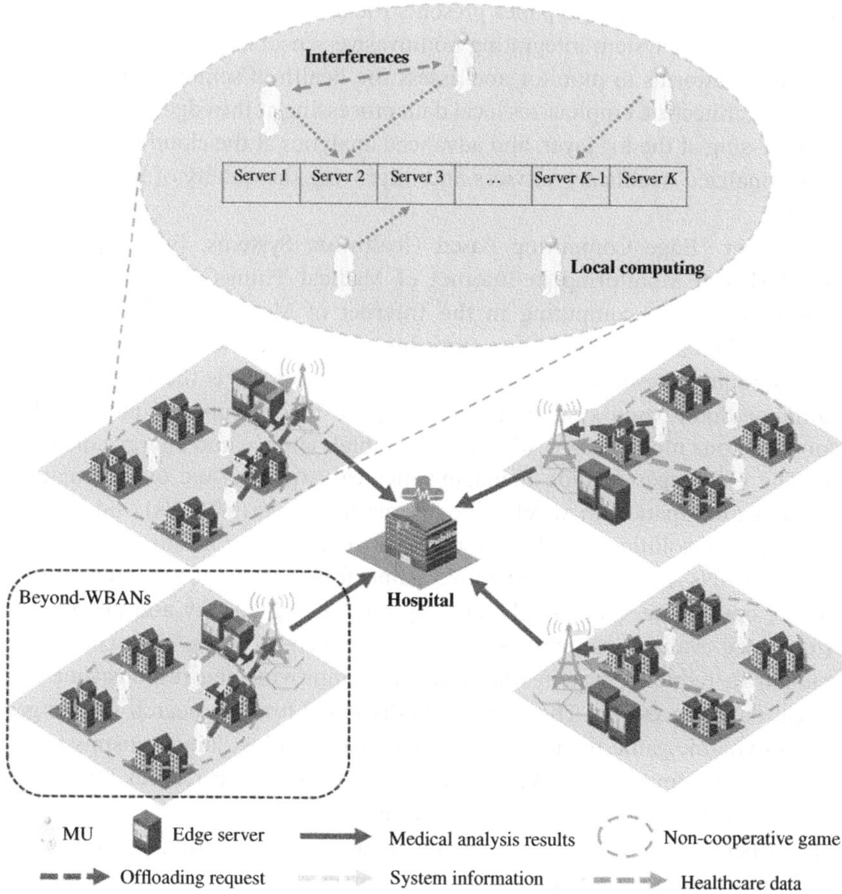

Figure 7.4 Edge computing in healthcare. Source: Dong et al. [18]/IEEE.

The paper "IoHT and Edge Computing, Warrants of Optimal Responsiveness of Monitoring Applications for Seniors: A Case Study" [22] investigates the integration of the Internet of Health Things (IoHT) and edge computing to enhance the monitoring and healthcare of senior patients. The primary problem addressed is the increasing burden on healthcare systems due to the aging population and the necessity for more efficient, real-time health monitoring solutions that empower seniors to manage their health independently. The challenges include handling large volumes of diverse, sensitive health data and ensuring low-latency, reliable data processing, and transmission. The solution proposed involves utilizing IoHT to gather extensive health and environmental data through connected devices and sensors, and leveraging edge computing to process this data close to its source. This approach reduces latency, enhances real-time responsiveness, and maintains data privacy. The paper presents the RO-SmartAgeing project, which aims to develop a system integrating noninvasive sensors, a smart environment, and cloud platforms to monitor and assess the health of seniors. The proposed six-layer architecture emphasizes local data processing at the edge, time-sensitive pre-processing at the fog layer, and advanced analytics at the cloud layer, providing personalized healthcare services and improving the quality of life for senior patients.

The paper "Edge Computing Based Healthcare Systems: Enabling Decentralized Health Monitoring in Internet of Medical Things" [18] explores the application of edge computing in the Internet of Medical Things (IoMT) to address challenges in healthcare monitoring, such as deficient wireless channel and computation resources. The main problem addressed is the inefficiency of traditional cloud-based systems in handling the high volume of medical data from numerous mobile users, which leads to high latency and energy consumption. The challenges include managing medical urgency, Age of Information, and energy dissipation in wireless body area networks (WBANs) and beyond. The proposed solution involves a cooperative game for resource allocation within WBANs and a noncooperative game for offloading decisions beyond WBANs, optimizing system-wide costs by minimizing latency and energy use. Performance evaluations demonstrate the effectiveness of the edge computing paradigm in reducing system-wide costs and improving the responsiveness of medical data processing. The paper also discusses future research challenges, such as dynamic game modeling using machine learning, joint transmission of energy and information, blockchain-based electronic health records for privacy preservation, and intelligent spectrum allocation.

The paper "Internet of Medical Things and Edge Computing for Improving Healthcare Systems in Smart Cities" [4] discusses the integration of edge computing with IoMT to enhance healthcare services. The problem addressed is the inefficiency of traditional cloud-based systems in handling the vast amounts of data generated by IoMT devices, leading to latency issues and increased costs. The paper identifies several challenges, including the need for real-time data processing, high energy consumption, and maintaining patient privacy and data security. The solution proposed involves leveraging edge computing to process data closer to the source, thereby reducing latency and bandwidth usage. Edge devices are used to perform initial data processing and analysis, only sending critical information to the cloud for further processing. This approach enhances the responsiveness of healthcare applications, reduces the load on centralized cloud servers, and ensures more efficient use of network resources.

The paper "Edge-assisted Healthcare Monitoring: Investigating the Role of Edge Computing in Real-time Monitoring and Management of Healthcare Data" by Ramswaroop Reddy Yellu et al. [47] explores the transformative impact of edge computing on healthcare monitoring systems. The primary problem addressed is the inefficiency of traditional centralized systems, which suffer from latency and potential privacy issues when processing vast amounts of healthcare data. The challenges include managing the complexity of numerous edge devices, ensuring data security and privacy, and achieving seamless integration with existing healthcare infrastructure. The solution proposed involves leveraging edge computing to process data closer to its source, such as medical devices and sensors, thereby reducing latency, enhancing data privacy, and improving scalability and reliability. The paper highlights several applications of edge-assisted healthcare monitoring, including remote patient monitoring, wearable health devices, telemedicine, emergency response systems, and predictive maintenance. Implementation strategies focus on selecting appropriate edge devices, optimizing data processing, ensuring robust data storage and security, and maintaining reliable network connectivity. Future research directions include integrating edge computing with AI and machine learning, addressing scalability and interoperability, and exploring regulatory and ethical considerations. The paper concludes that edge computing holds significant potential to revolutionize healthcare monitoring by enabling real-time data processing and improving patient outcomes.

7.4 Smart Cities

The integration of edge computing in smart city environments (as shown in Figure 7.5) offers numerous benefits and addresses critical challenges associated with big data analysis and IoT device management. Studies reveal that edge computing significantly reduces latency and bandwidth usage by processing data closer to its source, enhancing real-time decision-making and system efficiency. For instance, the Alternating Direction Method of Multipliers (ADMM) in edge servers alleviates the computational burden on central servers, enabling effective distributed data processing. Edge computing frameworks, like the one proposed for situational awareness, capture and process IoT data at the edge, sending only essential information to the cloud, which improves response times and situational accuracy. Real-time surveillance and traffic monitoring benefit from edge computing by reducing latency and enhancing anomaly detection through local data processing. However, challenges such as managing heterogeneous data sources, ensuring data privacy, and maintaining efficient resource allocation remain. Effective solutions involve multilayered architectures, adaptive fault-tolerant algorithms, and cooperative fog computing systems to balance computational workloads, energy efficiency, and fairness among nodes. Despite these challenges, edge computing demonstrates substantial potential to optimize

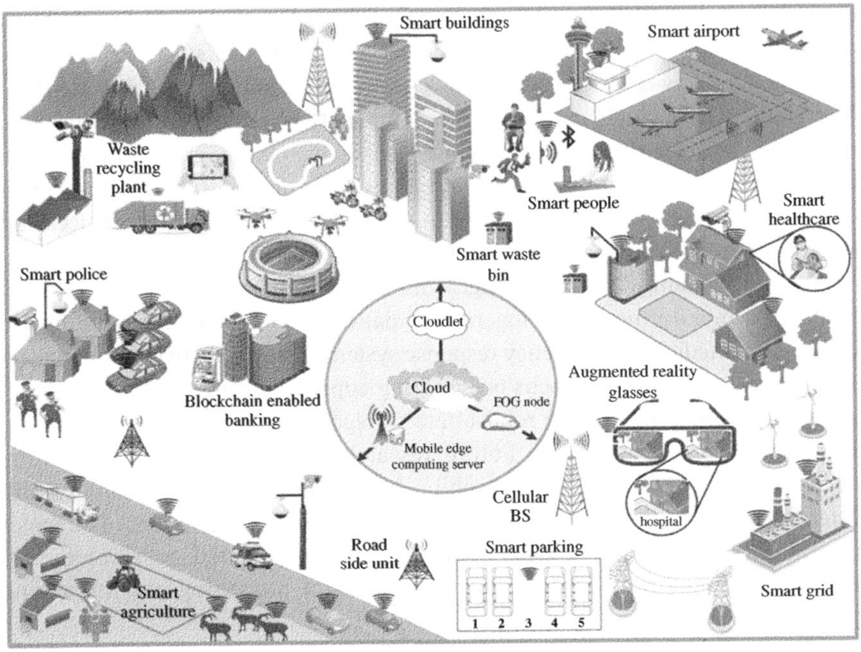

Figure 7.5 High-level view of an IoT-based smart city. Source: Khan et al. [25]/IEEE.

urban management, enhance energy efficiency, and improve public safety, making it a pivotal technology for the future of smart cities.

The paper "Incorporating Intelligence in Fog Computing for Big Data Analysis in Smart Cities" [41] in 2017 investigates the integration of edge computing and fog computing to address the challenges of big data analysis in smart city environments. The problem at hand is the inefficiency and high latency associated with centralized cloud computing for processing the vast amount of data generated by IoT devices in smart cities. Challenges include the need for low latency, efficient data distribution, and decentralized processing to manage the sheer volume of data streams from numerous mobile and static sensors. The proposed solution involves the use of ADMM for distributed data processing across edge servers, effectively reducing the computational burden on centralized cloud servers and enhancing performance. This approach allows for the splitting and parallel processing of data, improving overall system efficiency and enabling real-time data analysis essential for smart city applications. The study demonstrates the practical application of this method through dynamic naming conventions for mobile sensors, logical connections, and XML-based edge object structures to facilitate effective data management and processing.

In 2018, the paper "Edge Computing Framework for Enabling Situation Awareness in IoT Based Smart City" [1] by SK Alamgir Hossain et al. explore the implementation of an edge computing framework to enhance situation awareness in IoT-enabled smart cities. The primary problem addressed is the inefficiency of traditional cloud-based systems in processing the vast amounts of heterogeneous data generated by IoT devices, which leads to high latency and reduced performance. The challenges include handling data from diverse sources in real-time and ensuring low latency while maintaining data privacy. The proposed solution involves leveraging edge computing to process data closer to its source, thus minimizing latency and bandwidth usage. The framework captures raw IoT data, processes it at the edge, and sends only the necessary information to the cloud for further analysis and storage. This method provides situational awareness by generating "situation images" (S-images) that help decision-makers understand and respond to various urban conditions efficiently. The framework's effectiveness is demonstrated through experiments showing significant improvements in processing time and situation detection accuracy, highlighting the potential of edge computing in enhancing smart city operations.

Also in 2018, The paper "Smart City Surveillance at the Network Edge in the Era of IoT: Opportunities and Challenges" [10] by Ning Chen and Yu Chen explores the integration of fog computing to enhance surveillance in smart cities. The primary problem addressed is the difficulty in processing the vast amounts of data generated by ubiquitous sensors in urban environments, which traditional cloud computing methods struggle with due to high latency and bandwidth consumption. The challenges include efficiently detecting anomalies in real-time, ensuring timely decision-making, and managing the heterogeneity of data sources. The proposed solution leverages fog computing to process and store data closer to the source, thereby reducing latency and bandwidth usage. This approach allows for more effective real-time surveillance and quicker response times. The paper presents a case study on urban traffic surveillance, demonstrating how fog computing can improve anomaly detection and vehicle tracking in real-time. By distributing computational tasks across edge devices, the fog computing paradigm offers a scalable and efficient solution for the latency-sensitive applications required in smart city surveillance.

In 2020, the paper "Edge Computing with Big Data Cloud Architecture: A Case Study in Smart Building" by Catherine Inibhunu and Carolyn McGregor [23] presented an innovative framework for managing data in smart buildings using a combination of edge and cloud computing technologies. The core problem addressed is the complexity of efficiently handling vast amounts of environmental data generated by smart buildings, such as those used for climatic simulations at facilities like the ACE (Automotive Centre of Excellence) in

Ontario. The challenges include ensuring real-time data processing, maintaining data privacy and security, and integrating diverse data sources effectively. The proposed solution involves a multilayered architecture: Layer 1 handles data acquisition through sensors; Layer 2 uses edge computing nodes for local processing; and Layer 3 leverages cloud computing for advanced analytics and storage. This architecture facilitates real-time data processing, reduces latency, and improves scalability. The implementation is demonstrated through a prototype system at the ACE facility, which processes data from environmental simulations and supports various applications, including energy management and human-centered research. The paper underscores the potential of this integrated approach to enhance the functionality and efficiency of smart buildings while addressing data management and privacy challenges.

The paper titled "A Review on Edge Computing in Smart Energy by means of a Systematic Mapping Study" [40] explores the role of edge computing in enhancing smart energy systems. The primary problem addressed is the inefficiency of traditional centralized energy management systems in handling the massive data generated by smart energy devices, leading to high latency and energy consumption. The challenges include integrating heterogeneous data sources, ensuring real-time data processing, and maintaining security and privacy in distributed environments. The solution proposed involves using edge computing to process data closer to its source, thereby reducing latency, improving response times, and optimizing energy usage. The paper systematically maps existing research, categorizing studies based on publication type, research type, and the type of asset developed, such as architectures, frameworks, methods, or models. It identifies gaps in current research and suggests future directions for developing edge computing solutions to create more efficient, cost-effective, and real-time responsive smart energy systems.

The paper titled "Intelligent System Architecture for Smart City and its Applications Based Edge Computing" [2] discusses the integration of edge computing into smart city infrastructure to enhance the efficiency and responsiveness of urban services. The primary problem addressed is the limitation of traditional cloud computing in handling the vast amount of data generated by IoT devices in smart cities, which can lead to high latency and bandwidth issues. The challenges identified include managing the distributed and heterogeneous nature of edge devices, ensuring data security and privacy, and optimizing resource allocation to handle the dynamic and real-time requirements of smart city applications. The proposed solution involves a multi-layered architecture that leverages edge computing to process data closer to the source, thereby reducing latency and improving real-time decision-making capabilities. This architecture includes components for data acquisition, preprocessing, analytics, and service delivery, all designed to operate efficiently at the edge. The paper also highlights case studies in smart transportation and environmental monitoring, demonstrating the practical benefits of the proposed system in improving urban management and sustainability.

The paper "Edge Computing for IoT: A Use Case in Smart City Governance" [33] by Saurabh Nimkar and Dr. M.M. Khanapurkar in 2021 addresses the challenges of managing the vast data generated by IoT devices in smart cities, which traditional cloud computing struggles with due to latency and bandwidth issues. The core problem is efficiently handling and processing this data to support smart city applications, such as smart healthcare, energy management, and transportation. The main challenges include ensuring low latency, reducing network traffic, and maintaining data privacy and security. The proposed solution involves implementing a scalable architecture for IoT as a Service to Governance using edge computing. This architecture decentralizes data processing, bringing computational capabilities closer to the data sources, thus reducing latency and improving real-time responsiveness. The paper discusses various smart city applications where this architecture can be beneficial, such as smart grids, transportation systems, and waste management. The edge computing framework significantly enhances the efficiency and effectiveness of smart city governance by enabling real-time data processing and decision-making, thereby facilitating better service delivery to citizens.

The paper "Intelligent Edge Computing for IoT-Based Energy Management in Smart Cities" [29] by Yi Liu et al. explores the integration of edge computing with IoT for efficient energy management in smart cities. The main problem addressed is the challenge of managing the vast amounts of data generated by IoT devices in urban environments to optimize energy consumption. Traditional cloud-based systems face limitations due to high latency and bandwidth constraints. The proposed solution involves deploying an IoT-based energy management system with edge computing infrastructure enhanced by deep reinforcement learning (DRL). This system processes data locally at the network edge, reducing latency and transmission costs while enabling real-time decision-making. The paper presents a comprehensive framework and software model, along with an efficient energy scheduling scheme using DRL. Experimental results demonstrate that the proposed system significantly outperforms traditional cloud-based methods in terms of energy cost and delay, highlighting its potential to enhance energy management and sustainability in smart cities.

The paper "Edge-Computing Video Analytics for Real-Time Traffic Monitoring in a Smart City" [7] by Johan Barthélemy et al. presents an innovative edge-computing solution for real-time traffic monitoring using video analytics. The primary problem addressed is the inefficiency and high costs associated with traditional CCTV networks used for urban traffic monitoring, which often involve high bandwidth usage and privacy concerns. The proposed solution involves deploying smart visual sensors that leverage existing CCTV infrastructure and edge computing to process video data locally. This reduces bandwidth requirements and enhances privacy since only processed metadata, not raw video footage, is transmitted. The sensors employ you only look once (YOLO) V3 for object detection and simple online and realtime tracking (SORT) for tracking, ensuring real-time performance. The system's effectiveness is demonstrated through a pilot project in Liverpool, NSW, Australia, where sensors were used to monitor pedestrian and vehicle traffic, showing significant improvements in data-processing efficiency and real-time responsiveness. The paper concludes that integrating edge computing with IoT for smart city applications can optimize urban traffic management while addressing privacy and scalability challenges.

"An Edge Computing Based Public Vehicle System for Smart Transportation" [27] proposes an innovative solution to enhance public vehicle systems by leveraging edge computing to reduce latency and improve rider satisfaction and traffic efficiency. The primary problem addressed is the challenge of efficiently matching multiple riders to vehicles while considering personal preferences, which often leads to high computational overhead and long latency when handled by centralized data centers. Key challenges include managing the diverse preferences of travelers (e.g., travel time, distance, and costs) and ensuring real-time response to ride requests. The proposed solution, an edge computing-based public vehicle (ECPV) system, introduces an edge computing-based ride request transmission mechanism and a tree-based heuristic matching mechanism. These mechanisms enable efficient local processing of ride requests and dynamic vehicle scheduling, reducing decision-making delays and computational load on central servers. By utilizing unmanned ground vehicles (driverless cars) and strategically placing depots based on a graph partitioning method, the ECPV system achieves higher vehicle occupancy ratios, lower travel times, and reduced travel costs. Extensive simulations demonstrate the system's effectiveness in improving traveler satisfaction and overall traffic efficiency compared to traditional centralized approaches.

The paper "Edge Computing and Adaptive Fault-Tolerant Tracking Control Algorithm for Smart Buildings: A Case Study" [9] explores the integration of edge computing and advanced control algorithms to enhance energy efficiency and reliability in smart building environments. The primary problem addressed is the high energy consumption associated with temperature control in smart buildings and the frequent failures in control and monitoring systems. The main challenges include managing the large volume of heterogeneous data from IoT devices and ensuring robust and adaptive control under varying conditions and disturbances. The proposed solution involves a new adaptive control algorithm based on consensus game theory, which includes a state prediction module and a data quality module to improve the accuracy and reliability of IoT sensors and actuators. This algorithm optimizes temperature control, reduces tracking error, and enhances energy efficiency by predicting future states of precision and adjusting control actions accordingly. The effectiveness of the proposed system is demonstrated through a case study, showing significant improvements in maintaining desired temperature levels and reducing energy costs in a smart building setting.

The paper titled "Energy-Efficient Fair Cooperation Fog Computing in Mobile Edge Networks for Smart City" [17] by Yifan Dong et al. explores the challenges of managing high computational workloads and network latency in smart cities, especially with the integration of AI algorithms. The primary problem addressed is the need for an energy-efficient cooperation policy among fog nodes (FNs) to enhance the QoE for users while ensuring fairness among the nodes. The solution proposed involves constructing a cooperative fog computing system to process offloading workloads across the fog layer. The authors formulate a joint optimization problem that balances QoE and energy consumption while maintaining fairness among FNs. They prove the convexity of the optimization problem and design a Fairness Cooperation Algorithm (FCA) to obtain the optimal cooperation policy. Numerical results demonstrate that the FCA quickly converges and effectively reduces time overhead and energy consumption compared to traditional optimization approaches. The paper concludes that the proposed system outperforms existing algorithms in terms of time cost, energy efficiency, and fairness, making it a viable solution for the computational demands of smart cities.

The paper "Public Safety in Smart Cities under the Edge Computing Concept" [31] by Evangelos Maltezos et al. addresses the transformation of cities into smart cities to improve citizens' living conditions through modernized urban management and advanced technologies. The primary problem identified is the challenge of processing the vast amounts of data generated by compute-intensive security and safety applications in real time, which traditional cloud computing struggles to handle due to high latency and limited context awareness. The solution proposed involves the use of edge computing to process data closer to its source, thus enabling low-latency, context-aware, and geo-distributed capabilities. The paper introduces the Distributed Edge Computing IoT Platform (DECIoT), which integrates with smart building sensing systems and chemical precursor spectroscopic systems. DECIoT utilizes open-source microservices for data gathering, filtering, security, system management, and alert generation, ensuring efficient and secure processing of public safety data. This platform aims to enhance the safety and resilience of urban infrastructures and services by providing real-time, actionable insights and improving the efficiency of emergency responses in smart cities.

7.5 Internet of Things

Edge computing enhances IoT applications by reducing latency and improving data-processing efficiency by bringing computational resources closer to data sources. This approach significantly benefits real-time applications like mobile gaming and industrial IoT by ensuring low latency, reliable data transmission, and efficient handling of large data volumes. However, challenges include managing data offloading and load balancing due to the distributed nature of IoT devices, limited computing resources at the edge, and ensuring robust security and privacy. Addressing these challenges involves developing advanced load-balancing algorithms, optimizing 5G network integration, and employing technologies like blockchain for secure data sharing.

The paper "Edge Computing for the Internet of Things: A Case Study" [36] explores the necessity and benefits of edge computing in enhancing IoT applications, particularly through a case study on mobile gaming. The problem addressed is the high latency and unreliable communications often encountered in cloud-based IoT systems, which can degrade the user experience. The challenges highlighted include the need for low latency, reliable data transmission, and efficient processing of large volumes of sensor data. The solution proposed involves leveraging edge computing architectures, which bring computational resources closer to the data sources, thereby reducing latency and improving data-processing efficiency. The paper classifies various edge computing platforms and demonstrates, through an experimental evaluation, that edge computing significantly enhances the quality of experience for mobile gaming applications. This approach can be extended to other IoT applications requiring similar real-time data-processing capabilities, showcasing the broader potential of edge computing in the IoT ecosystem.

The paper "Edge Computing in Industrial Internet of Things: Architecture, Advances and Challenges" [37] explores the integration of edge computing into the Industrial Internet of Things (IIoT) to enhance efficiency and performance. The primary challenge addressed is the latency in decision-making processes, which is crucial for real-time applications in IIoT. The authors propose an edge computing architecture to decentralize data processing, thereby reducing latency and bandwidth consumption while enhancing privacy and security. Key challenges include efficient data offloading and load balancing due to the distributed nature of IIoT devices, and the limited computing resources of edge devices. The paper discusses full and partial data offloading schemes to optimize these processes and highlights the potential of combining edge artificial intelligence to handle complex computations. Additionally, the paper emphasizes the importance of secure data sharing in edge computing environments, suggesting the use of blockchain technology for enhanced security. Future research directions include improving network slicing for 5G integration, optimizing load balancing algorithms, and developing robust security frameworks to address the unique challenges of IIoT edge computing.

The paper "Learning IoT in Edge: Deep Learning for the Internet of Things with Edge Computing" [26] explores the integration of deep learning with edge computing to enhance the processing capabilities of IoT applications. The primary challenge addressed is the inefficiency of centralized cloud computing in handling the vast amounts of data generated by IoT devices, which often results in high latency and bandwidth issues. The proposed solution involves deploying deep learning models partially on edge servers, closer to the data sources, to reduce the amount of data transmitted to the cloud and improve processing efficiency. The paper presents a novel scheduling strategy to optimize the deployment of deep learning tasks in an edge computing environment, maximizing the number of tasks that can be handled while ensuring quality of service. Experimental results demonstrate that this approach significantly outperforms traditional methods, enhancing both the performance and scalability of IoT applications. The paper concludes by suggesting future work in deploying these strategies in real-world environments to further validate their effectiveness.

7.6 Retail

Edge computing is revolutionizing customer experiences and operational efficiency (as shown in Figure 7.6). Retailers implement edge computing to manage inventory smarter and personalize customer shopping experiences. For instance, smart shelves equipped with weight sensors and cameras can track inventory levels in real time, automatically signaling replenishment needs. Additionally, edge computing enables the analysis of customer traffic and buying patterns directly at the store level, allowing for immediate promotional adjustments and personalized customer interactions. This localized data processing helps retailers respond more dynamically to consumer behavior, enhancing customer satisfaction and increasing sales. The paper "Intelligent Communication Between IoT Devices on Edges in Retail Sector" [39] proposes the Smart Shop (SmSH) architecture designed to enhance the shopping experience by integrating IoT devices and edge computing in the retail sector. The SmSH architecture leverages technologies like RFID, NFC, Bluetooth, and low-power wide-area network (LPWAN) to facilitate seamless communication between IoT devices, enabling real-time data processing and personalized customer interactions. By processing data at the edge, the architecture reduces latency, enhances security, and improves the efficiency of store operations, such as inventory management and customer assistance. The paper discusses the implementation phases of SmSH, including

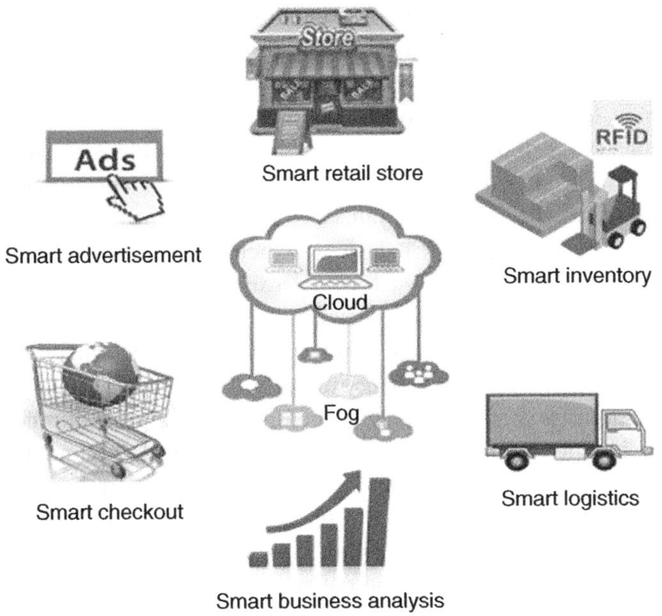

Figure 7.6 High-level view of an IoT-based smart retail. Source: Perera et al. [34]/ ACM, Inc.

customer identification, product selection, and payment, while highlighting the role of edge computing in ensuring fast and secure data handling. Additionally, the architecture's adaptability allows for easy extension to other business domains, making it a versatile solution for modern retail challenges.

The paper "Survey on Multi-Access Edge Computing for Internet of Things Realization" [35] provides a comprehensive overview of how multi-access edge computing (MEC) can enhance the realization of IoT applications. It discusses the significant role of MEC in extending cloud computing capabilities to the edge of the network, thereby reducing latency, improving bandwidth utilization, and enhancing the overall performance of IoT systems. The survey highlights key application scenarios such as smart homes, healthcare, autonomous vehicles, and industrial IoT, where MEC can offer substantial benefits by processing data closer to the source, ensuring real-time operations, and improving scalability. However, the integration of MEC with IoT also presents challenges, including managing scalability, ensuring reliable communication, efficient computation offloading, and addressing security, privacy, and trust issues. The paper concludes by summarizing the state-of-the-art technologies and future research directions necessary for the successful integration of MEC in IoT environments, emphasizing the importance of continued innovation to overcome these challenges and fully realize the potential of MEC in the IoT ecosystem.

7.7 Autonomous Vehicles

Edge computing significantly enhances the performance of connected and autonomous vehicles (CAVs) by reducing latency and improving data-processing efficiency (as shown in Figure 7.7). Frameworks like OpenVDAP and VECFrame address the limitations of onboard computation by offloading intensive tasks to nearby edge servers, optimizing resource utilization, and ensuring low-latency communication. However, challenges such as managing computational overhead, ensuring data privacy and security, and handling large volumes of sensor data persist. Collaborative learning and federated learning frameworks, such as CLONE, leverage edge computing to maintain data privacy while improving prediction accuracy. Solutions like SafeCross and collaborative autonomous driving framework (CCAD) enhance vehicle safety and traffic management by providing real-time warnings and improving perception capabilities through edge-based data processing. Despite these advancements, the field faces ongoing challenges in efficient resource allocation, energy consumption, and network

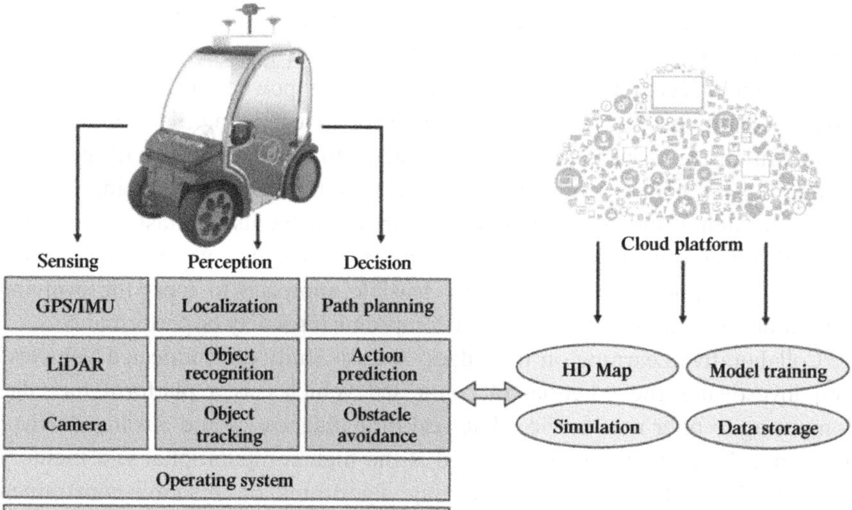

Figure 7.7 High-level view of edge computing for autonomous vehicles. Source: Liu et al. [28]/IEEE.

stability, necessitating continued research and development to fully realize the potential of edge computing in CAVs.

"OpenVDAP: An Open Vehicular Data Analytics Platform for CAVs" [50] presents a comprehensive edge computing framework designed to address the computational and latency challenges faced by CAVs. The primary problem tackled is the inefficiency of onboard computation in handling data-intensive services such as real-time diagnostics, advanced driver-assistance systems, and third-party applications due to limited computing resources. Challenges include ensuring low latency, managing computational overhead, and preserving security and privacy in a dynamic vehicular environment. OpenVDAP offers a solution by leveraging a full-stack edge computing platform that includes an onboard heterogeneous computing unit, a secure and privacy-preserved operating system (EdgeOSv), and an edge-aware application library (libvdap). The framework dynamically offloads computational tasks to nearby edge servers or the cloud, optimizing resource utilization and reducing latency. OpenVDAP's open-source nature encourages community collaboration, allowing researchers to deploy and evaluate applications in real-world settings, thus enhancing the performance and reliability of CAV systems.

"Vehicular and Edge Computing for Emerging Connected and Autonomous Vehicle Applications" [6] addresses the complexities and computational demands of CAVs. The problem focuses on the need for optimal computing resource allocations and efficient architectures to handle the real-time data from advanced sensors such as cameras, radars, and LiDARs. Challenges include ensuring low latency, high accuracy, reliability, and managing power consumption under varying conditions. The solution proposed involves leveraging edge computing to offload intensive tasks from in-vehicle systems to roadside units with powerful computing capabilities, thereby enhancing performance and reducing latency. The paper demonstrates through preliminary experiments that task partitioning and offloading strategies can significantly improve the efficiency and effectiveness of vehicular applications, providing a feasible approach to meet the evolving computational needs of CAVs without compromising safety or performance.

"Collaborative Learning on the Edges: A Case Study on Connected Vehicles" [30] investigates the potential of connected vehicles as a platform for edge computing to offer new services like real-time diagnostics and advanced driver assistance. The key problem addressed is the high computational and memory resource demands of machine learning algorithms on resource-constrained edge devices. Challenges include handling large datasets for training, preserving privacy, and reducing latency while maintaining security. The solution proposed is CLONE, a collaborative learning framework based on federated learning and long short-term memory networks. This approach allows vehicles to train models locally with their data and share parameter updates with a central edge server, enhancing prediction accuracy while reducing training time and maintaining data privacy. The case study on predicting electric vehicle battery failures demonstrated the efficacy of this method, highlighting improved prediction accuracy by including driver behavior metrics and the superior performance of LSTMs over other models like random forests and gradient-boosting decision trees.

"VECFrame: A Vehicular Edge Computing Framework for Connected Autonomous Vehicles" [42] presents a novel framework aimed at addressing the limitations of current autonomous vehicle systems that rely heavily on cloud computing. The problem identified is the high latency and network congestion associated with cloud-based solutions, which hinder real-time object detection and data processing essential for autonomous driving. The key challenges include the efficient transfer and fusion of large volumes of sensor data from multiple vehicles, ensuring scalability, adaptability, and maintaining low-latency communication. VECFrame proposes a solution by leveraging edge computing to perform cooperative object detection and data fusion at the edge of the network, closer to the vehicles. This framework utilizes modular containers for data dissemination, enabling a scalable and platform-independent approach. Real-world experiments demonstrate that VECFrame significantly improves the accuracy of traffic condition perception, enhances object detection capabilities, and increases data throughput by 40%–350% compared to nonedge solutions, thus providing a robust and efficient infrastructure for connected and autonomous vehicles.

"Offloading Autonomous Driving Services via Edge Computing" [13] addresses the critical challenge of processing massive amounts of sensor data in real time to ensure safe and reliable decisions in autonomous driving. The main problem is the insufficient onboard computational resources of autonomous vehicles to meet these demands. The paper proposes a novel solution involving the offloading of computationally intensive tasks to roadside units and the cloud, thereby leveraging edge computing. This approach utilizes an integer linear programming formulation for offline optimization of the scheduling strategy and a fast heuristics algorithm for online adaptation. Experimental results from both synthetic task graphs and real-world deployments show significant improvements in system performance, including reduced average latency by 34% and enhanced localization accuracy by up to 7.6 times. The proposed method effectively balances the computational load between onboard units, edge servers, and cloud resources, addressing the challenges of bandwidth limitations and network instability.

"Vehicle Selection and Resource Optimization for Federated Learning in Vehicular Edge Computing" [46] addresses the efficient processing of data in vehicular edge computing (VEC) using federated learning (FL). The main problem tackled is the significant energy consumption and time required for model training and transmission, compounded by the variability in vehicles' computational capabilities and data quality. The key challenges include selecting appropriate vehicles for FL tasks and optimizing resource allocation under the constraints of learning time and energy consumption. The paper proposes a min–max optimization approach that dynamically selects vehicles based on their data quality and computational capabilities while minimizing overall system cost through a greedy algorithm. This involves decomposing the optimization problem into two subproblems: resource allocation, solved using the Lagrangian dual method and subgradient projection, and local model accuracy optimization, addressed with an adaptive harmony search algorithm. Simulations demonstrate that the proposed algorithms effectively balance cost and resource optimization, achieving significant performance improvements in FL for VEC scenarios.

"To Turn or Not To Turn, SafeCross is the Answer" [45] addresses the significant challenge of blind areas in left-turn scenarios at intersections, which pose a considerable threat to driver safety and can lead to fatal collisions. Despite advancements in vision-based perception technologies that enable autonomous driving systems to achieve a 360-degree view and avoid most blind areas, the issue persists when an opposing road is blocked by another vehicle at the intersection. To tackle this problem, the authors propose SafeCross, a framework designed to monitor intersections and provide real-time warnings to left-turning vehicles when another vehicle is detected in the blind area. The framework comprises four main components: video preprocessing, video classification, few-shot learning, and model switching. These components work together to train a model that identifies blind areas, adapts to different weather conditions, and provides accurate warnings. Experimental results demonstrate that SafeCross enhances vehicle safety and increases left-turn traffic throughput by 50%, highlighting its effectiveness in mitigating blind area risks.

"FAIR: Towards Impartial Resource Allocation for Intelligent Vehicles with Automotive Edge Computing" [44] explores the challenges and solutions related to resource allocation in intelligent vehicle systems using edge computing. The main problem addressed is the inefficient and often biased allocation of computational resources, which can hinder the performance and safety of intelligent vehicles. The paper highlights the challenges of ensuring fair resource distribution in environments with diverse and competing demands. It proposes a novel framework, FAIR (Fair and Impartial Resource Allocation), which leverages edge computing to optimize the allocation process. The solution involves advanced algorithms that dynamically adjust resource distribution based on real-time data and predefined fairness criteria, ensuring that all vehicles receive adequate computational power for critical tasks such as navigation, obstacle detection, and communication. The proposed framework is validated through extensive simulations and real-world experiments, demonstrating significant improvements in both resource utilization efficiency and the overall performance of intelligent vehicular networks.

"Towards C-V2X Enabled Collaborative Autonomous Driving" [21] investigates the limitations of single-agent autonomous vehicles, which rely solely on their onboard sensors, leading to frequent accidents due to restricted sensing coverage. The primary problem identified is the insufficient safety and reliability of current autonomous systems operating independently. The key challenges include limited perception angles, missed detection of safety-critical information, and difficulties in lane-keeping under adverse conditions. The authors propose a C-V2X-enabled CCAD, which employs cellular vehicle-to-everything (C-V2X) technology to enable communication between vehicles and infrastructure, thus enhancing the perception capabilities through multiple angles. The framework incorporates edge computing for real-time data processing and has demonstrated significant improvements in lane-keeping accuracy and overall vehicle safety through a case study. This collaborative approach effectively addresses the sensing limitations of single-agent systems, providing a robust solution for safer autonomous driving.

7.8 Summary and Practice

7.8.1 Summary

This chapter focuses on the transformative impact of edge computing across various industries, including manufacturing, IoT, retail, healthcare, telecommunications, autonomous vehicles, and smart cities. Edge computing, by processing data closer to its source, enhances operational efficiency, decision-making, and innovation. In manufacturing, it enables predictive maintenance and real-time quality control. For IoT, it reduces latency and improves system efficiency and security. In retail, it aids in personalized services and inventory management. Healthcare benefits from real-time patient monitoring and diagnostics. Telecommunications leverage edge computing to optimize network performance and support emerging technologies like 5G and VR. Autonomous vehicles rely on it for real-time decision-making, while smart cities utilize it for efficient resource management and enhanced public safety. Through detailed case studies, this chapter elucidates the practical applications, challenges, and benefits of edge computing, underscoring its critical role in modern technological advancements and industry improvements.

7.8.2 Practice Questions

1) How does edge computing improve operational efficiency in the manufacturing sector?
2) What are the key benefits of edge computing for IoT ecosystems, particularly in terms of latency and data security?

3) Describe the role of edge computing in enhancing real-time patient monitoring and diagnostics in healthcare.
4) What are the primary challenges of implementing edge computing in autonomous vehicles, and how are they addressed?
5) Discuss the impact of edge computing on public safety and resource management in smart cities.

7.8.3 Course Projects

1) Analyze and present a case study of a successful edge computing implementation in a specific industry.
2) Develop a predictive maintenance system for manufacturing using edge computing, and evaluate its performance in reducing downtime and improving productivity.
3) Investigate the integration of edge computing with IoT devices in a healthcare setting, focusing on real-time patient monitoring and data security.
4) Create a simulation model to compare the performance of edge computing vs. cloud computing in processing IoT data for autonomous vehicles.
5) Research and analyze the role of edge computing in enhancing 5G network performance and supporting emerging applications like AR and VR.

Chapter 7 Suggested Papers

1 Latif U Khan et al. "Edge-computing-enabled smart cities: A comprehensive survey". In: *IEEE Internet of Things Journal* 7. 10 (2020), pp. 10200–10232.
2 Sidi Lu, Yongtao Yao, and Weisong Shi. "Collaborative learning on the edges: A case study on connected vehicles". In: *2nd USENIX Workshop on Hot Topics in Edge Computing (HotEdge 19)*, 2019.
3 Shaoshan Liu et al. "Edge computing for autonomous driving: Opportunities and challenges". In: *Proceedings of the IEEE* 107. 8 (2019), pp. 1697–1716.

References

1 S K Alamgir Hossain, Md Anisur Rahman, and M Anwar Hossain. "Edge computing framework for enabling situation awareness in IoT based smart city". In: *Journal of Parallel and Distributed Computing* 122 (2018), pp. 226–237.
2 Mehdhar Al-gaashani et al. "Intelligent system architecture for smart city and its applications based edge computing". In: *2020 12th International Congress on Ultra Modern Telecommunications and Control Systems and Workshops (ICUMT)*. IEEE, 2020, pp. 269–274.

3 Ali Alnoman. "Edge computing services for smart cities: A review and case study". In: *2021 International Symposium on Networks, Computers and Communications (ISNCC)*. IEEE, 2021, pp. 1–6.

4 Muna Alrazgan. "Internet of medical things and edge computing for improving healthcare in smart cities". In: *Mathematical Problems in Engineering* 2022. 1 (2022), p. 5776954.

5 Mohammed Alrowaily and Zhuo Lu. "Secure edge computing in IoT systems: Review and case studies". In: *2018 IEEE/ACM Symposium on Edge Computing (SEC)*. IEEE, 2018, pp. 440–444.

6 Sabur Baidya et al. "Vehicular and edge computing for emerging connected and autonomous vehicle applications". In: *2020 57th ACM/IEEE Design Automation Conference (DAC)*. IEEE, 2020, pp. 1–6.

7 Johan Barthélemy et al. "Edge-computing video analytics for real-time traffic monitoring in a smart city". In: *Sensors* 19. 9 (2019), p. 2048.

8 Abhiraj Biswas, Ayush Jain, and Mohana. "Survey on edge computing–key technology in retail industry". In: *Computer Networks and Inventive Communication Technologies: Proceedings of the 3rd ICCNCT 2020*. Springer, 2021, pp. 97–106.

9 Roberto Casado-Vara et al. "Edge computing and adaptive fault-tolerant tracking control algorithm for smart buildings: A case study". In: *Cybernetics and Systems* 51. 7 (2020), pp. 685–697.

10 Ning Chen and Yu Chen. "Smart city surveillance at the network edge in the era of IoT: Opportunities and challenges". In: *Smart Cities: Development and Governance Frameworks* (2018), pp. 153–176.

11 Baotong Chen et al. "Edge computing in IoT-based manufacturing". In: *IEEE Communications Magazine* 56. 9 (2018), pp. 103–109.

12 Gianfranco Ciccarella et al. "Edge cloud computing in telecommunications: Case studies on performance improvement and TCO saving". In: *2019 4th International Conference on Fog and Mobile Edge Computing (FMEC)*. IEEE, 2019, pp. 113–120.

13 Mingyue Cui et al. "Offloading autonomous driving services via edge computing". In: *IEEE Internet of Things Journal* 7. 10 (2020), pp. 10535–10547.

14 Thiago Pereira Da Silva et al. "Fog computing platforms for smart city applications: A survey". In: *ACM Transactions on Internet Technology* 22. 4 (2022), pp. 1–32.

15 Sujata Dash et al. "Edge and fog computing in healthcare–A review". In: *Scalable Computing: Practice and Experience* 20. 2 (2019), pp. 191–206.

16 Rushit Dave, Naeem Seliya, and Nyle Siddiqui. "The benefits of edge computing in healthcare, smart cities, and IoT". In: *arXiv preprint arXiv:2112.01250* (2021).

17 Yifan Dong et al. "Energy-efficient fair cooperation fog computing in mobile edge networks for smart city". In: *IEEE Internet of Things Journal* 6. 5 (2019), pp. 7543–7554.

18 Peiran Dong et al. "Edge computing based healthcare systems: Enabling decentralized health monitoring in Internet of medical Things". In: *IEEE Network* 34. 5 (2020), pp. 254–261.

19 Mohammed S Elbamby et al. "Wireless edge computing with latency and reliability guarantees". In: *Proceedings of the IEEE* 107. 8 (2019), pp. 1717–1737.

20 Shahid Sultan Hajam and Shabir Ahmad Sofi. "IoT-Fog architectures in smart city applications: A survey". In: *China Communications* 18. 11 (2021), pp. 117–140.

21 Y He et al. "Towards C-V2X enabled collaborative autonomous driving". In: *IEEE Transactions on Vehicular Technology* 72. 12 (2023), pp. 15450–15462. DOI: 10.1109/TVT.2023.3299844.

22 Marilena Ianculescu et al. "IoHT and edge computing, warrants of optimal responsiveness of monitoring applications for seniors. A case study". In: *2019 22nd International Conference on Control Systems and Computer Science (CSCS)*. IEEE, 2019, pp. 655–661.

23 Catherine Inibhunu and Carolyn McGregor. "Edge computing with big data cloud architecture: A case study in smart building". In: *2020 IEEE International Conference on Big Data (Big Data)*. IEEE, 2020, pp. 3387–3393.

24 Balaram Yadav Kasula. "The role of edge computing in real-time analytics for smart city healthcare applications". In: *Transaction on Recent Developments in Industrial IoT* 9. 9 (2017), pp. 1–7.

25 Latif U Khan et al. "Edge-computing-enabled smart cities: A comprehensive survey". In: *IEEE Internet of Things Journal* 7. 10 (2020), pp. 10200–10232.

26 He Li, Kaoru Ota, and Mianxiong Dong. "Learning IoT in edge: Deep learning for the Internet of Things with edge computing". In: *IEEE Network* 32. 1 (2018), pp. 96–101.

27 J Lin et al. "An edge computing based public vehicle system for smart transportation". In: *IEEE Transactions on Vehicular Technology* 69. 11 (2020), pp. 12635–12651. DOI: 10.1109/TVT.2020.3028497.

28 Shaoshan Liu et al. "Edge computing for autonomous driving: Opportunities and challenges". In: *Proceedings of the IEEE* 107. 8 (2019), pp. 1697–1716.

29 Yi Liu et al. "Intelligent edge computing for IoT-based energy management in smart cities". In: *IEEE Network* 33. 2 (2019), pp. 111–117.

30 Sidi Lu, Yongtao Yao, and Weisong Shi. "Collaborative learning on the edges: A case study on connected vehicles". In: *2nd USENIX Workshop on Hot Topics in Edge Computing (HotEdge 19)*, 2019.

31 E Maltezos et al. “Public safety in smart cities under the edge computing concept”. In: *2021 IEEE International Mediterranean Conference on Communications and Networking (MeditCom)*. 2021, pp. 88–93. DOI: 10.1109/MeditCom49071.2021.9647550.

32 Garima Nain, K K Pattanaik, and G K Sharma. “Towards edge computing in intelligent manufacturing: Past, present and future”. In: *Journal of Manufacturing Systems* 62 (2022), pp. 588–611.

33 Saurabh Nimkar and M M Khanapurkar. “Edge computing for IoT: A use case in smart city governance”. In: *2021 International Conference on Computational Intelligence and Computing Applications (ICCICA)*. IEEE, 2021, pp. 1–5.

34 Charith Perera et al. “Fog computing for sustainable smart cities: A survey”. In: *ACM Computing Surveys (CSUR)* 50. 3 (2017), pp. 1–43.

35 Pawani Porambage et al. “Survey on multi-access edge computing for Internet of Things realization”. In: *IEEE Communications Surveys & Tutorials* 20. 4 (2018), pp. 2961–2991.

36 Gopika Premsankar, Mario Di Francesco, and Tarik Taleb. “Edge computing for the Internet of Things: A case study”. In: *IEEE Internet of Things Journal* 5. 2 (2018), pp. 1275–1284.

37 Tie Qiu et al. “Edge computing in Industrial Internet of Things: Architecture, advances and challenges”. In: *IEEE Communications Surveys & Tutorials* 22. 4 (2020), pp. 2462–2488.

38 Partha Pratim Ray, Dinesh Dash, and Debashis De. “Edge computing for Internet of Things: A survey, e-healthcare case study and future direction”. In: *Journal of Network and Computer Applications* 140 (2019), pp. 1–22.

39 M Saravanan and N C Srinidhi Srivatsan. “Intelligent communication between IoT devices on edges in retail sector”. In: *Advances in Information and Communication Networks: Proceedings of the 2018 Future of Information and Communication Conference (FICC), Vol.* 2. Springer, 2019, pp. 546–562.

40 Inés Sittón-Candanedo et al. “A review on edge computing in smart energy by means of a systematic mapping study”. In: *Electronics* 9. 1 (2019), p. 48.

41 Bo Tang et al. “Incorporating intelligence in fog computing for big data analysis in smart cities”. In: *IEEE Transactions on Industrial Informatics* 13. 5 (2017), pp. 2140–2150.

42 S Tang et al. “VECFrame: A vehicular edge computing framework for connected autonomous vehicles”. In: *2021 IEEE International Conference on Edge Computing (EDGE)*. 2021, pp. 68–77. DOI: 10.1109/EDGE53862.2021.00019.

43 Jumyung Um et al. “Edge computing in smart production”. In: *Advances in Service and Industrial Robotics: Proceedings of the 28th International Conference on Robotics in Alpe-Adria-Danube Region (RAAD 2019) 28*. Springer, 2020, pp. 144–152.

44 H Wang, J Xie, and M M A Muslam. “FAIR: Towards impartial resource allocation for intelligent vehicles with automotive edge computing”. In:

IEEE Transactions on Intelligent Vehicles 8. 2 (2023), pp. 1971–1982. DOI: 10.1109/TIV.2023.3234888.

45 B Wu et al. "To turn or not to turn, safecross is the answer". In: *2022 IEEE 42nd International Conference on Distributed Computing Systems (ICDCS)*. 2022, pp. 414–424. DOI: 10.1109/ICDCS54860.2022.00047.

46 H Xiao et al. "Vehicle selection and resource optimization for federated learning in vehicular edge computing". In: *IEEE Transactions on Intelligent Transportation Systems* 23. 8 (2022), pp. 11073–11087. DOI: 10.1109/TITS.2021.3099597.

47 Ramswaroop Reddy Yellu, Praveen Thuniki, and Mohan Raparthi. "Edge-assisted Healthcare Monitoring: Investigating the role of edge computing in real-time monitoring and management of healthcare data". In: *African Journal of Artificial Intelligence and Sustainable Development* 4. 1 (2024), pp. 70–78.

48 Wenjin Yu et al. "Edge computing-assisted IoT framework with an autoencoder for fault detection in manufacturing predictive maintenance". In: *IEEE Transactions on Industrial Informatics* 19. 4 (2022), pp. 5701–5710.

49 Chaoyang Zhang and Weixi Ji. "Edge computing enabled production anomalies detection and energy-efficient production decision approach for discrete manufacturing workshops". In: *IEEE Access* 8 (2020), pp. 158197–158207.

50 Qingyang Zhang et al. "OpenVDAP: An open vehicular data analytics platform for CAVs". In: *2018 IEEE 38th International Conference on Distributed Computing Systems (ICDCS)*. IEEE, 2018, pp. 1310–1320.

51 Yongli Zhao et al. "Edge computing and networking: A survey on infrastructures and applications". In: *IEEE Access* 7 (2019), pp. 101213–101230.

8

Privacy and Bias in Edge Computing*

As an emerging computing paradigm, edge computing has made a significant impact on modern computing systems and modern applications, introducing a range of ethical and social implications. In this chapter, we will explore two key topics, privacy and bias, which are closely intertwined with these ethical and social concerns.

The first major ethical and social consideration is privacy, which becomes a critical issue whenever data containing sensitive information is shared by its owner in any type of applications. In edge computing scenarios, edge server process large amounts of data offloaded from end devices, raising substantial privacy concerns. However, edge computing also offers a new avenue and unique opportunity to design, implement, and deploy privacy-preserving mechanisms that avoid sending sensitive data to more powerful and more publicly accessible cloud servers. In this chapter, we will present a set of major privacy concerns associated with edge computing and explore various privacy-preserving solutions.

The second ethical and social consideration is bias. Due to the localized nature of edge computing, data available at the a single edge node may be significant biased, which can lead to potentially misleading conclusions when they are applied in other contexts. Furthermore, edge computing often relies on various machine learning (ML) and deep learning algorithms for autonomous decision-making, and these algorithms could be inherently biased. In this chapter, we will analyze various types of biases, their causes, their impact on edge computing algorithms, and techniques to mitigate the effects of these biases.

*This chapter is contributed by Kewei Sha, Komala Subramanyam Cherukuri, and Haihua Chen.

Edge Computing: Systems and Applications, First Edition. Lanyu Xu and Weisong Shi.

Companion Website: www.wiley.com/go/xu/computing

8.1 Privacy in Edge Computing

Modern applications, such as large-scale IoT systems (IoT), connected autonomous vehicles, and smart communities, generate, collect, and analyze vast amounts of data. Edge computing has emerged as a critical paradigm that extends close services to support the efficient performance of these applications. Unlike Cloud computing, edge computing processes data closer to its sources, significantly enhancing efficiency by reducing communication overhead and delay.

Privacy ensures freedom from unauthorized surveillance and documentation, requiring service providers in digital environments to handle personal data responsibly, safeguard identities, and keep actions untraceable [54]. While edge computing inherently reduces privacy concerns compared to cloud computing, since the data does not need to be transferred and processed at a centralized cloud, it still presents significant privacy concerns. This is particularly true given the sensitive nature of the data involving personal information, such as user identities, locations, activities, and behaviors [59]. These privacy concerns can deter users from fully adopting, utilizing, and benefiting from advanced applications. Therefore, it is essential to design and implement privacy-preserving techniques that address these concerns. Edge computing, being proximate to the data source and having sufficient resources to process a huge amount of data, is ideally suited for deploying such privacy-preserving mechanisms [60].

In Sections 8.1.1 and 8.1.4, we first explore the specific privacy concerns associated with edge computing and then present various types of privacy. Finally, we will discuss a set of effective privacy-preserving solutions and highlight several existing edge-based implementations.

8.1.1 Privacy Concerns at Edge Computing

The rapid advancement of technology, coupled with the widespread use of big data and the Internet of Things (IoT), has led to significant privacy breaches impacting both individuals and businesses [3]. Information privacy, defined as the right to control how personal data is collected and utilized, enables individuals or groups to prevent their personal information from being disclosed without their consent. One of the major privacy concerns today is the risk of personal data being identified during its transmission, storage, and analysis, leading to unauthorized access and misuse [27]. The implementation of regulations like the European General Data Protection Regulation [25] and the ongoing digitization of various sectors have highlighted the importance of data anonymization in data processing and analysis [56].

In edge computing, where user data is often outsourced to third-party edge services, there is a clear separation between data ownership and control. This separation increases the risk of data loss, leakage, and unauthorized data operations, underscoring the need for strong data security and privacy-preserving mechanisms [54]. Additionally, the architectural design of edge computing, which consists of multiple layers, including cloud, edge, and end devices, introduces several new privacy concerns. Below is a list of these concerns.

- **Distributed data storage**: Unlike centralized systems, edge computing distributes data across multiple locations. This segmentation complicates the assurance of data integrity and increases the risk of privacy breaches, such as data leakage or unauthorized modifications across different edge nodes.
- **Retention of identifying information**: Untrusted servers, providers, and operators at the edge may retain user-identifiable information even after service relationships have ended. This retention practice exacerbates privacy concerns, highlighting the difficulty of securing outsourced data in decentralized environments [4].
- **Vulnerability to honest But curious adversaries**: In edge computing, sensitive information transmitted to edge servers is vulnerable to privacy breaches. This vulnerability is particularly concerning given the presence of "honest but curious" adversaries, authorized entities that exploit their access to gather sensitive information for personal interests.
- **Predictability of user locations**: Because edge computing brings processing closer to the data source, users' location data often becomes more predictable. This predictability necessitates stringent privacy protections to prevent unauthorized tracking and exploitation.
- **Decentralized edge administration**: The decentralized and dispersed nature of edge devices complicates centralized control, increasing the network's vulnerability to cyberattacks. These attacks can lead to privacy breaches, data theft, and system hijacking, causing widespread disruption. If an edge node is compromised, attackers can exploit these vulnerabilities to access and misuse sensitive personal information.
- **Privacy risk in data transmission**: Recent research highlights the privacy concerns that arise during the collaboration between cloud, edge, and end devices [59], particularly when raw data such as texts and images are transmitted. This process introduces the risk of data leakage during transmission, which can compromise sensitive information.
- **Uncontrollable data sharing**: Once data is transferred to edge servers, the original data owner loses control over its sharing. Entities providing edge services may share consumer data with external organizations for purposes such as behavior analysis and market expansion, thereby increasing privacy risks.

8.1.2 Various Forms of Privacy

The discussion above presents privacy concerns in edge computing-related systems and applications. These concerns affect various aspects of privacy, such as data privacy, location privacy, identity privacy, and communication privacy. Among these, data privacy, location privacy, and identity privacy are currently major ones in the edge computing assisted applications. In the following text, we define each of these key privacy forms.

8.1.2.1 Data Privacy

Data privacy in the context of edge computing refers to the protection of sensitive information that is stored, processed, or transmitted across decentralized edge servers and end devices. Ensuring the user data against privacy leaks is essential, particularly as edge computing supports big data applications in sectors like healthcare, banking, and crowd-sourcing, where large volumes of personal data are generated and require stringent privacy measures [54]. For example, as shown in the left side of Figure 8.1, smart home security systems use cameras to monitor user movements and activities; Personal finance apps track sensitive financial data; and voice assistant software may continually record and analyze voice data. Ensuring data privacy is crucial due to the challenges posed by the distributed nature of edge servers, which are harder to monitor than centralized data centers [17]. The frequent transfer of data between devices and edge servers increases the risk of exposure, especially when data is migrated across different servers to optimize user experience.

8.1.2.2 Location Privacy

Location privacy is a significant concern as numerous web services and applications require users to share their location to access certain functionalities. As indicated in the middle of Figure 8.1, examples include public transportation systems apps that collect user locations; ride-sharing platform that tracks user movements; and fitness tracker that provides many location-aware services. The potential leakage of this sensitive information poses a real and considerable risk to users [7]. The proximity of end devices to edge servers not only facilitates task offloading for optimized performance but also significantly risks location privacy, as curious edge servers can infer the proximity of end users from their connection points. As users move and their devices switch between servers, the communication paths can reveal their movement patterns to unauthorized servers, compromising their location privacy [17]. Users might unintentionally or unknowingly consent to share their location through pop-up requests, without fully understanding the potential consequences, enabling diverse applications, posing significant risks by potentially exposing users' geo-locations. This exposure can threaten financial, entertainment, professional, and confidential aspects of life, leading to risks such as hijacking, blackmail, or ransom [54].

8.1.2.3 Identity Privacy

Identity privacy in edge computing refers to the protection of personal information during interactions that require identity verification, such as filling out online

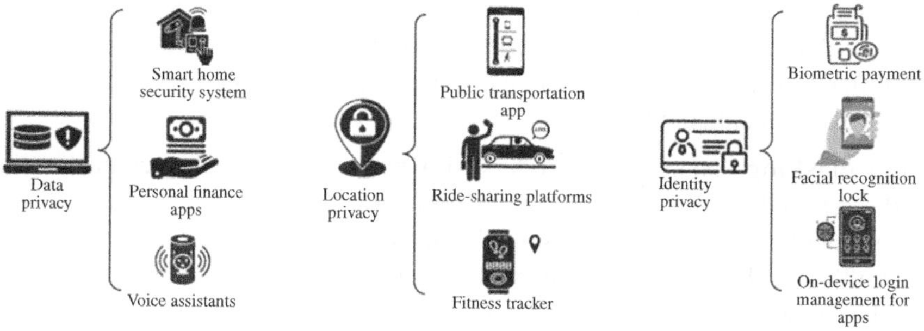

Figure 8.1 Privacy forms.

forms, where information stored on edge servers is accessed for authorization purposes, linking to sensitive details like payment data [17]. With the rise of tactile Internet and IoT, safeguarding identity becomes crucial as it involves billions of entities and people, using methods such as knowledge-based (username), possession-based (random number generator), inherence-based (biometric like facial recognition), or physical unclonable function (PUF) for identity verification, as indicated in the right side of Figure 8.1. An adversary who replicates a user identity could access all associated data, posing risks of privacy violations through tampering, cloning, and masquerading [54].

8.1.3 Introduction of Privacy-Preserving Techniques

To tackle privacy concerns in edge computing, this section surveys several important methods designed to protect data, location, and identity privacy, with an emphasis on techniques such as k-anonymity, differential privacy, federated learning, and homomorphic encryption. These methods help ensure that sensitive information is safeguarded against unauthorized access and potential data breaches. Each technique offers a unique approach to enhancing privacy, collectively providing robust defenses against privacy risks in edge computing environments.

8.1.3.1 Data Anonymization

Data anonymization is a practical and widely used solution for protecting user privacy in data publishing. It involves the removal of any information that can uniquely, directly, or indirectly identify someone from the data [38]. User identifiers include direct identifiers, such as names, addresses, and photos, which can directly identify a user, and indirect identifiers, such as ages, salaries, and occupations, which can identify a user by linking with other available datasets or information [55], such information is removed from the data before its anonymization [38]. Many edge computing assisted applications collect extensive data to improve their services and develop new products, but this practice significantly increases the risk of data loss or accidental data breaches [55]. Various anonymization techniques have been proposed and implemented for different scenarios, making anonymization a practical solution for preserving user privacy in data publishing [38].

Data masking: Data masking refers to a method of changing characters in the selected attribute(s) to a different character, making the variable inconceivable [44]. This involves using character modification techniques such as shuffling, substitution, and encryption to hide data [55]. For example, email address "John.Doe123@example.com" and apply the described data masking technique, the modified email address would be "Zzzz.Zzz111@zxxxxzz.zzz."

Generalization: Generalization is the process of replacing specific values with less-specific but semantically consistent values [44], often by altering data into a set of ranges, to make it unidentifiable [55]. This technique applies at the cell level, where some original values are maintained with added confusion, thereby increasing the difficulty for an attacker to infer sensitive data [44]. For example, the original data showing "John Smith, 29, US$ 50,000" can be generalized to "Person 1, 20–30, US$ 40,000–US$ 60,000." This replaces specific values with ranges, making the data less identifiable while maintaining semantic consistency.

Suppression: Suppression refers to the removal of an entire part of data, such as a column or tuple, in a dataset by changing the value to one that does not have meaning [44]. This should occur whenever an attribute is irrelevant, unnecessary for analysis, or impossible to anonymize in any other way [40]. An example is replacing the original data with "****." Suppression conceals information by deleting it, ensuring that records in the original data are completely removed from the final output. This technique can be applied to columns or tuples to hide them when needed [44].

8.1.3.2 *k*-Anonymity

Privacy is a critical issue in edge computing, comparable in importance to other areas. The *k*-anonymity model is widely utilized for data privacy protection [37], as it helps prevent record linkage when data is released for public health or demographic research, ensuring the privacy of data subjects, involves removing certain identifiers [3].

k-Anonymity is a technique to prevent link attacks by generalizing and/or suppressing certain elements. This ensures that no individual is uniquely identifiable within a group of size k [35]. A dataset with *k*-anonymity ensures that each record is indistinguishable from at least $k - 1$ other records in the dataset. The larger the value of k, the higher the level of privacy, as it becomes more difficult to identify any individual with a probability greater than $1/k$ through linking attacks [11]. Increasing the *k*-value enhances the level of privacy protection [3]. k is a parameter that represents the level of anonymity in a dataset.

Despite the numerous anonymization methods available, *k*-anonymity remains popular due to its straightforward yet effective privacy guarantees. It employs techniques like suppression and generalization to obscure data [64]. Suppression involves deleting records from the dataset to protect privacy, while generalization involves replacing quasi-identifiers—attributes that could be used to reidentify individuals when combined with external data—with more general data. The challenge with *k*-anonymization lies in balancing data utility and privacy protection. The process can impact the usefulness of the data, as both data suppression and generalization can alter its utility. The challenge with *k*-anonymization lies in balancing data utility and privacy protection. The process can impact the usefulness of the data, as both data suppression and generalization can alter its utility. Therefore, achieving optimal solutions in *k*-anonymity is crucial, aiming to protect data privacy while minimizing the effects on data utility [37].

To better illustrate the concept of *k*-anonymity, an example is provided in Tables 8.1 and 8.2. Table 8.1 presents the original dataset for disease analysis, which contains sensitive patient information such as name, age, and zipcode. These details could be exploited by others to directly or indirectly identify individuals. To anonymize the data, operations, including suppressing names,

Table 8.1 Original dataset.

Name	Age	Gender	Zip code	Disease
Dana	24	F	45011	Diabetes
John	30	M	45012	None
Jack	28	M	45013	Asthma
Alex	26	M	45014	Hypertension
Chloe	27	F	45015	None
Fiona	29	F	45016	Diabetes

Table 8.2 Anonymized dataset with 2-anonymity.

Name	Age	Gender	Zip code	Disease
Suppressed	20–25	F	450**	Diabetes
Suppressed	26–31	M	450**	None
Suppressed	26–31	M	450**	Asthma
Suppressed	26–31	M	450**	Hypertension
Suppressed	26–31	F	450**	None
Suppressed	26–31	F	450**	Diabetes
Suppressed	20–24	F	450**	Diabetes

generalizing ages and zip codes, and adding a new entry at the end of the table, are performed. These modifications result in a dataset that satisfies 2-anonymity, meaning each record is indistinguishable from at least one other record in the dataset in terms of age, gender, and zip code. Please note this process achieves the goal of anonymization, but adding an additional entry may impact the disease analysis accuracy.

The above example shows how to preserve identity privacy, but k-anonymity can be applied to protect various forms of privacy, including identity, location, behavior, and data privacy in edge computing. For example, recent research has developed a k-anonymity mechanism to preserve behavior privacy by ensuring similar task offloading frequencies among groups of at least k users. This technique creates equivalence classes that obscure individual user behaviors [29]. By generalizing task frequencies within these groups, privacy protection is achieved. Adaptive strategies further refine this approach by adjusting offloading behaviors based on privacy thresholds, effectively managing privacy risks by switching between local and edge computing.

In another recent effort [76], researchers have devised a k-anonymity algorithm to protect location privacy for location-aware services. This system employs a method called dual k-anonymity, which hides user's real location by mixing it with several fake locations before transmitting it to the service provider. An intermediary server, known as an edge server, facilitates this process by handling the communication and also storing frequently requested location information. This approach not only makes it difficult for others to trace user's actual location but also improves service efficiency by using previously cached data for similar future requests, ensuring both privacy and performance.

8.1.3.3 Differential Privacy

Differential privacy is a technique designed to protect the privacy of individual records within statistical databases by ensuring that the output of a function is not sensitive to any specific record in the dataset. This approach helps to minimize the risks of identifying individual records, focusing on the privacy of these records rather than the dataset as a whole [69]. Differential privacy is a rigorous and provably secure method that does not depend on the background knowledge of attackers, effectively resisting membership inference attacks [31]. It is a notable privacy-preserving mechanism that maximizes the protection of data against adversaries with substantial background knowledge [74]. Therefore, differential privacy has gained significant attention and has been the subject of extensive research, especially in applications related to edge computing [13].

$$P[M(D) \in S] \leq e^{\epsilon} \cdot P[M(D') \in S] \tag{8.1}$$

Equation 8.1 shows the concept of ϵ-differential privacy, where, M is a random algorithm that provides ϵ-differential privacy, D and D' are neighboring datasets that differ by at most one record, S is any subset of the range of possible outputs, and ϵ is the privacy budget. This definition ensures that the presence or absence of any individual record in the dataset does not significantly affect the output of the algorithm, providing a quantifiable level of privacy [74].

8.1.3.4 Privacy Budget

The privacy budget in differential privacy refers to the allowable amount of noise that can be added to the data before it significantly impacts the model's accuracy. Balancing this budget is crucial to maintaining both privacy and data utility [7]. Privacy budget, often denoted as ϵ, controls the amount of noise added, balancing privacy protection and data utility [69].

8.1.3.5 Global Sensitivity

The sensitivity metric quantifies the extent of required perturbation within a differentially private mechanism. In a similar vein, global sensitivity is concerned with the maximum deviation between query outputs across two datasets that differ in only a single element (commonly referred to as neighboring datasets). For a randomized query $f : B \rightarrow \mathbb{R}$, the global sensitivity $\Delta_{f_{gs}}$ is determined using the following expression:

$$\Delta_{f_{gs}} = \max_{B,B'} \|f(B) - f(B')\| \tag{8.2}$$

The discourse on differential privacy can be categorized into two principal branches: existing differential privacy methods and noise addition mechanisms [23].

8.1.3.6 Noise Addition

Random noise in datasets obscures individual data points without compromising overall accuracy, thus protecting dataset privacy to a great extent [7]. The volume of noise added for privacy protection is unrelated to the dataset's size [69]. Instead, the noise addition is proportional to the variance in the dataset, and the noise added to satisfy differential privacy is based solely on the value of global sensitivity for the Sum query, independent of the actual dataset [57].

8.1.3.7 The Laplacian Mechanism

The Laplacian mechanism is a widely utilized approach in ϵ-differential privacy for query functions f that yield responses $f(D) \in \mathbb{R}^p$, wherein the concept of sensitivity is pivotal. Given a query function f and a norm function $\|\cdot\|$ over the range of f, the *sensitivity* $s(f, \|\cdot\|)$ is defined as:

$$s(f, \|\cdot\|) = \max_{d(D,D')=1} \|f(D) - f(D')\| \tag{8.3}$$

Typically, the norm function $\|\cdot\|$ is either the L_1 or L_2 norm. L_1, also known as the Manhattan norm, represents the sum of the absolute values of the differences, and L_2 represents the Euclidean norm, which is the square root of the sum of the squares of the differences.

The Laplacian mechanism operates by taking a query function f and a norm function over its range, producing a perturbed function $\hat{f}(D) = f(D) + \eta$ that adheres to ϵ-differential privacy. Here, η is a stochastic variable with a probability density function given by:

$$p(\eta) \propto e^{-\epsilon\|\eta\|/s(f,\|\cdot\|)} \tag{8.4}$$

Moreover, an alternative to the Laplacian mechanism involves substituting Laplacian noise (i.e., η in the above formula) with Gaussian noise. This modification significantly diminishes the probability of generating excessively large noise values; however, it retains (ϵ, δ)-differential privacy for some $\delta > 0$, which is a relaxation of the stricter ϵ-differential privacy criterion [28].

Recent research [13] explores the use of differential privacy to protect location privacy in edge computing. The core concept involves end devices performing perturbation, meaning they alter their location data before transmitting it to edge servers. This perturbation ensures that the data received by edge servers is sufficiently "noisy" to prevent the identification of specific users, while still maintaining overall statistical utility.

Specifically, the geographical area is divided into sections using Voronoi diagrams, with each cell in the diagram representing a distinct area of the network's edge. End devices independently perturb their location data in accordance with the principles of location differential privacy (LDP) before transmitting it to the nearest edge node. These edge nodes then gather the perturbed data points and perform necessary analyses or aggregate the data for further processing.

By applying LDP, the exact locations of users are obfuscated through the introduction of random noise, thereby significantly reducing the risk of privacy breaches. Additionally, because the data is processed close to its point of generation—at the edge of the network—the system can efficiently manage large volumes of data with minimal latency.

8.1.3.8 Homomorphic Encryption

Homomorphic encryption is a specialized encryption scheme that allows third parties to operate on encrypted data without decrypting it in advance [4]. This capability protects sensitive information by enabling data to be processed in an encrypted form, ensuring that only encrypted data is accessible to service providers [43].

Homomorphic encryption is advantageous as it allows operations on ciphertexts, yielding the same results as operations conducted directly on raw data [75]. Homomorphic encryption enables operations on plaintexts without decryption, allowing additions and multiplications on ciphertexts to reflect directly on the corresponding plaintexts. This capability allows for data manipulation within the encrypted domain [39]. As a result, homomorphic encryption is seen as a key approach to solving database query problems on encrypted data. With the increasing requirements for the privacy of digital data and the algorithms used to process more complex structures, homomorphic encryption aligns with the growth of communication networks, equipment, and their capacities [68].

Edge servers present potential risks as they can be compromised in certain situations, raising concerns about data privacy and security when transmitting raw data. Traditional encryption methods protect data only during transmission, leaving stored data vulnerable to unauthorized access or breaches. Although homomorphic encryption allows secure ciphertext processing, it typically incurs significant latency, rendering it impractical for real-time applications [10]. The process of signing data to ensure its integrity and authenticity can consume significant computing resources and lead to challenges such as high latency and unsafe data storage and sharing [75].

An encryption scheme is called *homomorphic* over an operation "$*$" if it supports the following equation:

$$E(m_1) * E(m_2) = E(m_1 * m_2), \quad \forall m_1, m_2 \in M \tag{8.5}$$

In Equation 8.5, E is the encryption algorithm and M is the set of all possible messages [4]. Homomorphic encryption enables operations on plaintexts without decryption, allowing additions and multiplications on ciphertexts to reflect directly on the corresponding plaintexts.

In homomorphic encryption, the "magic" lies in the ability to perform operations directly on encrypted data (ciphertext) to obtain a result that, when the result is decrypted, matches the result of performing these same operations on the original, unencrypted data. Figure 8.2 illustrates this concept with an example.

In the figure, the original data consists of two plaintext words, "Alex" and "Doctor." These words are separately encrypted using a specific method with public keys, resulting in ciphertexts, assumed to be "Dohc" and "Grfwru." Then, we perform an operation, represented by "+," on the ciphertext, and

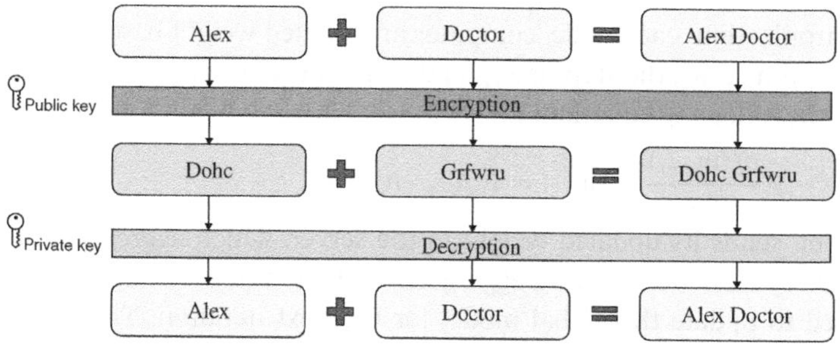

Figure 8.2 An example of homomorphic encryption.

producing an encrypted result, assumed to be "Dohc Grfwru." When this result is decrypted using private keys, the outcome is "Alex Doctor," identical to the result that would have been obtained if the operations had been performed on the original plaintext. This process ensures that the results are derived without exposing the original plaintexts, "Alex" and "Doctor."

When applied in edge computing, different end devices can locally encrypt their sensitive information and then send it to an edge server. Upon receiving encrypted data from multiple edge devices, the edge server can perform computationally intensive operations to analyze the encrypted data. Then the result will be sent back to each individual end device, where it can be decrypted and viewed in plaintext. In this process, the edge server's computational power is leveraged to perform complex operations while preserving data privacy.

However, one concern with homomorphic encryption is its significant consumption of computing resources, which can lead to high latency.

8.1.3.9 Federated Learning

FL is a ML technique that develops a global model using data from decentralized clients without the need to centralize the data [16]. This approach allows multiple distributed nodes to collaboratively train a shared prediction model using their local data [67]. Recent advancements in FL have significantly enhanced the practical application of secure multiparty computing theory. It preserves the privacy of source data by only interacting with the server through parameter updates. Additionally, FL aligns closely with edge computing principles, making their combination a prominent trend [31]. This process involves building a shared global model at an aggregator, such as a multi-access edge computing server located at a base station or an access point (AP) [46].

FL typically trains neural networks in a decentralized manner across numerous devices as shown in Figure 8.3. In this framework, a global neural network is maintained on a central server, while the training data for the neural network is distributed across multiple nodes (e.g., autonomous cars in the figure), often exhibiting heterogeneity. Consider several nodes, $\text{node}_1, \text{node}_2, \ldots, \text{node}_n$. Each node_i retains its private dataset ξ_i, and a local neural network's loss function, denoted by $f(\cdot)$.

During synchronization, each node_i computes an updated weight based on the current weight at time t, w_t^i, the step size γ_t, and its private dataset ξ_i. The update rule is given by the following equation:

$$w_{t+1}^i = w_t^i - \gamma_t \frac{\partial f(w_t, \xi_i)}{\partial w} \quad \text{for} \quad i = 1, 2, \ldots, n \tag{8.6}$$

Each node then sends its updated weight to the server, which aggregates the weights uploaded by all nodes using an aggregation function $A(\cdot)$. The aggregated weights are used to update the global model for the next iteration. The global

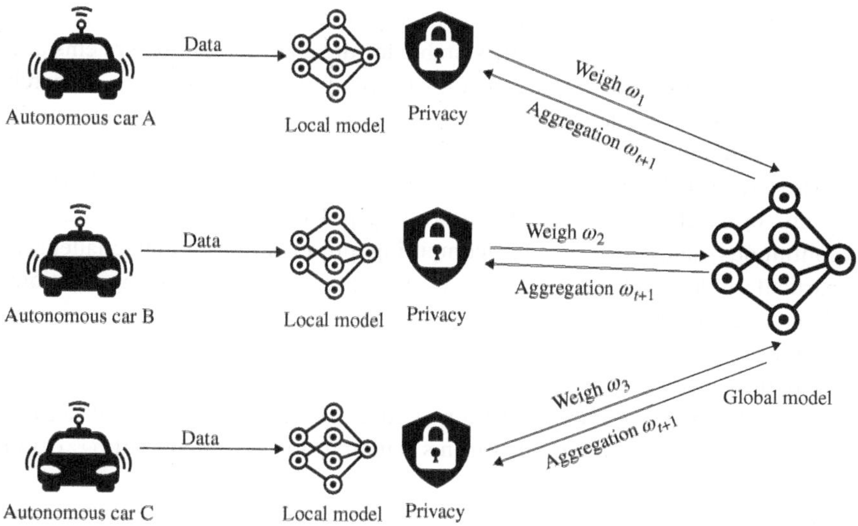

Figure 8.3 Federated learning.

model weights at time $t + 1$ are thus:

$$w_{t+1} = A(w^1_{t+1}, w^2_{t+1}, \dots, w^n_{t+1}) \tag{8.7}$$

The server sends the weights of the updated global model back to each node to trigger another iteration of local training. This process continues until a stop condition is met [30].

FL reduces communication costs, enhances data privacy, and improves system scalability [2, 70], while it also requires that end nodes should possess sufficient computational complexity to perform local model training, which could be a challenge for many [16].

Recent work [62] demonstrates how edge FL is applied for data analytics in autonomous vehicle networks. Each autonomous vehicle manages data from various sensors, including GPS (global positioning system), LiDAR (Light Detection and Ranging), and multiple onboard cameras, which are essential for intelligent navigation and early warnings [67]. The vehicles develop local models and collaborate with the edge server to train an effective global model following the steps described earlier. This approach enhances smart vehicle decision-making while preserving data privacy and maintaining efficiency.

8.1.4 Open Research Problems

Despite the development of numerous privacy-persevering solutions proposed in edge computing, there are still many open research problems. In the following text, we highlight some of these key areas.

- **Federated learning**: While FL has been proven to be an effective approach for preserving privacy, especially suitable for the edge computing systems and applications, it usually requires significant computational resources. Research is needed to find a balance between computational efficiency and privacy in FL for edge computing.
- **Differential privacy**: As a rigorous and provable method, differential privacy offers an excellent approach for protecting privacy; however, applying it to continuous data flow in edge computing presents substantial challenges, particularly for applications with real-time requirements. Exploring solutions that address these challenges is critical.
- **Homomorphic encryption**: Homomorphic encryption is a powerful tool for both security and privacy, but it is often computationally heavy, making it difficult for many resource-constrained edge devices. Developing lightweight homomorphic encryption algorithms tailored for edge computing could significantly enhance its practicality.
- **AI-driven anonymization**: Recent advances in artificial intelligence (AI) provide new opportunities to augment and anonymize the data without significantly compromising data utility. Exploring approaches based on transformer models and large language models could open new avenues for privacy preservation in edge computing.
- **Security, privacy, and trust trade-offs**: There are inherent trade-offs between security, privacy, and trust in collaborative edge computing environments, which usually involve multiple untrusted or semi-trusted parties. Research is needed to develop mechanisms that can establish and maintain trust and security while ensuring privacy across heterogeneous and untrusted edge nodes.
- **Legal and ethical regulations**: The legal and ethical implications of privacy in edge computing are critical for many edge computing systems and applications. We need to investigate various legal frameworks to regulate and ensure compliance and protect user rights.

8.2 Accessibility and Digital Divide

Bias is another important ethical and social concern in edge computing. This section provides an overview of the various types of biases, their causes, their impact on edge computing algorithms, and strategies for mitigating these biases.

8.2.1 What Is Bias?

Bias simply refers to deviation from a standard [19]. Bias in the context of algorithmic systems refers to a deviation from a normative standard. This term encompasses a variety of system behaviors, such as "gender bias" or "racial bias," which may impact different groups in diverse and harmful ways [14]. The concern with fairness and bias is integral to discussions of data justice, highlighting the potential of "big data" and algorithmic decision-making to exacerbate existing societal injustices [24]. Automated decision-making systems, even without malicious intent, can lead to unfair outcomes that disproportionately disadvantage or advantage specific groups defined by sensitive attributes like race or gender [72]. Bias can appear at multiple stages within the ML lifecycle, including data collection, preparation, modeling, evaluation, and deployment [5]. The increasing reliance on algorithms in various social and economic domains has prompted concerns about their potential to inadvertently perpetuate discrimination [33].

Edge deployments are inherently heterogeneous, with varying sensing capabilities and environmental conditions across different deployments. This heterogeneity disrupts the independence and identical distribution property of local data across distributed edge nodes, leading to biased global models. Such models contribute to unfair decision-making and discrimination against specific communities or groups [58], affecting system reliability, which is the ability of on-device ML to consistently deliver stable and predictable performance under expected operating conditions [26].

8.2.2 Types of Biases

Biases can be examined from various perspectives, leading to many different types. In the following text, we introduce a list of frequently encountered biases. We also identified one specific cause and corresponding mitigation technique for each type of bias.

Data bias: Data bias refers to the inherent variability and potential prejudices present in datasets. These biases can emerge due to the diverse quality of data sources [9]. This widespread data bias significantly influences the algorithms we develop to enhance user experiences, affecting their accuracy and fairness [8]. For instance, individuals working in government, universities, and other information-centric institutions typically produce data of higher quality and reduced bias. In contrast, social media platforms, with their extensive and diverse user base, tend to generate more biased and lower-quality data [9].

Algorithmic bias: Algorithmic bias occurs when an algorithm introduces bias that was not present in the input data. Algorithms can both reflect and amplify human or structural biases, or create their own complex biases [65]. If an algorithm is trained on biased data, it is likely to reinforce the patterns from the dominant category within that data [47]. This bias can exist even if the algorithm's developer did not intend to discriminate [22]. It arises from the differential usage of information in the input or training data, leading to biased outputs [19].

While the input data might be unbiased, the algorithm itself can still generate bias. Defining a fair algorithmic approach is complex, often requiring human expertise to determine if an output is biased [9]. Furthermore, algorithms can create new biases, such as presentation bias, which alters future interaction data, thereby generating additional biases [8].

Here we provide two examples of algorithm bias. In this first example, even when the training data is diverse, the algorithm is designed to prioritize accuracy across the entire dataset rather than ensuring equal performance across different demographic groups. As a result, the algorithm might work exceptionally well for identifying faces of certain demographics (e.g., middle-aged white males) but performs poorly on others (e.g., women, people of color, or older adults). In another example, consider an algorithm trained on a dataset with 80% healthy and 20% diseased images. This algorithm could achieve 80% accuracy by simply classifying all samples as healthy, even though it fails to correctly identify diseased cases [47]. Such an algorithm would be significantly biased when applied to disease analysis.

Cause: Algorithmic bias in ML-based models often stems from inaccurate datasets that misrepresent the target demographics, inadequate model attributes, or inaccurate development and deployment methods [6]. This bias can manifest through the differential treatment of information within the input data, often exacerbated when models are inappropriately used or deployed across different contexts, leading to biased outcomes [19].

Mitigation: Algorithmic bias often arises when models are overfitted to noisy or atypical data. To mitigate this, many algorithms incorporate smoothing or regularization techniques that help prevent such overfitting [19]. Ideally, the development of new algorithms that inherently lack such biases would be a preferable solution. Additionally, personalization algorithms, by focusing on individual user relevance rather than aggregate popularity metrics, can reduce biases linked to widespread preferences [15]. However, developing ML applications that are unbiased requires a comprehensive skill set, encompassing data collection, integration, algorithm development, and oversight during algorithm training [6].

Statistical bias: Statistical bias occurs when an algorithm produces results that deviate from the true underlying estimate. This type of bias is prevalent in predictive algorithms due to various factors such as suboptimal sampling methods, measurement errors in predictor variables, and the heterogeneity of effects [51]. In statistical terms, bias arises when the distribution of a dataset does not accurately represent the true distribution of the population, leading the algorithm to produce outputs that differ from the actual estimate [47]. An example of statistical bias is when a medical AI system trained primarily on data from young adults fails to accurately predict health outcomes for older adults. This happens because the data does not represent the entire population, leading to skewed predictions that do not account for the differences in age groups.

Sampling bias: Sampling bias is a type of representation bias that occurs from nonrandom sampling of specific subgroups. Sampling bias causes the patterns observed in one group to be inaccurately generalized to new groups from different populations [41]. For instance, if a survey about job satisfaction is conducted exclusively within high-tech industries, the result might misleadingly suggest higher job satisfaction levels across all industries. This occurs because other sectors, such as retail or healthcare, are not included. Although sample bias shares similarities with statistical bias, they are different types of bias. Sample bias arises during the data collection, that is, how the sample is selected, while statistical bias could occur at various stages, including data collection, model selection, etc.

Social bias: Social bias arises when external influences, such as the behavior of others, skew our judgments [65]. This type of bias can result from algorithmic inaccuracies, which might be statistically biased, or from human elements, such as implicit or explicit biases [51]. An example of social bias, a job application system prioritizes resumes that include certain elite universities, it may unintentionally favor candidates from more privileged backgrounds, overlooking equally talented individuals from less recognized schools.

Historical bias: Historical bias reflects preexisting societal and technical biases that can infiltrate data, even with ideal sampling and feature selection processes [65]. This bias often affects automated decision-making systems that rely on historical data for training [72]. An example of an automated hiring system uses past employment records to make decisions. If these past records show a trend of hiring more men than women for certain roles, the system might continue to favor men, perpetuating the existing gender imbalance.

Representation bias: Representation bias occurs due to the sampling methods used when collecting data from a population [65]. Representation bias occurs when the training data inadequately represents certain segments of the target population, leading to poor generalization to those groups. For instance, a survey aimed at assessing illegal drug use among teenagers may be biased if it only includes responses from high school students while excluding home-schooled students or dropouts [61].

Group bias: Group bias refers to the discrepancy in classification performance among different groups based on a shared global model in FL. A group is defined as a subset of data with samples categorized by a specific sensitive attribute such as race, gender, class, or label. In this context, group bias occurs when the model's performance varies significantly between these groups [58]. For example, consider a predictive model used across hospitals: it might perform well for urban hospitals but poorly for rural ones due to differences in patient demographics and regional health trends, reflecting a clear group bias.

Attribute bias: Attribute bias occurs in the undirected attributed network when the value distributions of certain attributes differ between demographic groups based on a sensitive attribute. This bias can develop even if the original attributes are unbiased, as it can emerge after attributes are propagated through the network. This type of bias is identified by examining whether the information propagation in the network introduces or exacerbates discrepancies in attribute distributions across demographic groups [20]. For example, in a social network, if users primarily share information about their college degrees within their community, the network may propagate this information unevenly, creating an illusion that a higher percentage of the community holds college degrees. This distorted view can lead to incorrect assumptions about the education level of the entire community, even though the original data may not have been biased.

Structural bias: Structural bias in an undirected attributed network exists when the propagated attribute values show different distributions between demographic groups at any attribute dimension. This type of bias indicates that the network structure or information propagation process creates or amplifies differences between groups based on sensitive attributes, resulting in biased outcomes [20]. An example of structural bias in an undirected attributed network occurs when a social media platform connects users and shares job ads differently among groups. If the platform's structure leads to some racial groups mostly seeing lower-paying job ads while others see higher-paying ads, this indicates a structural bias. It shows that the network's design itself might unfairly distribute opportunities based on race.

Measurement bias: Measurement bias, also known as reporting bias, occurs due to the selection, utilization, and measurement of specific features [65]. For example, a healthcare study when self-reported data is used to assess exercise levels. If participants overestimate their activity due to social desirability or poor recall, the results may not accurately reflect true behaviors, skewing conclusions about exercise interventions. This introduces measurement bias, compromising the study's validity.

Evaluation bias: Evaluation bias happens during model evaluation [65]. For example, if an AI health diagnostic tool is evaluated using data mainly from middle-aged individuals, it may not perform well for younger or older populations. This shows population bias because the evaluation did not consider all relevant age groups.

Population bias: Population bias emerges when the user population of a platform exhibits different statistics, demographics, representatives, and characteristics compared to the intended target population [65]. For example, in a health app study, the data gathered on exercise habits mainly comes from younger, tech-savvy users because they are more likely to use such digital tools. This creates a population bias as it fails to capture the exercise patterns of older adults, who may be less inclined to use technology.

8.2.3 Causes of Biases?

Bias in ML can originate from several sources [52], making it a multifaceted issue [20] that affects the fairness and accuracy of AI systems [8]. One major source of bias is the datasets used for training, which often contain inherent biases due to various reasons such as biased device measurements, historical human decision biases, erroneous reports, or other factors. These inherent biases in data are compounded by algorithmic biases that arise from the objective of minimizing overall aggregated prediction errors, which typically favors majority groups over minority ones [52].

The fairness perspective further highlights that the homophily of sensitive attributes, such as race, gender, or nationality, directly influences predictions and introduces inequalities [63]. Sources of bias in datasets include missing values, small sample sizes, and misclassification or measurement errors [21]. These issues lead to datasets that do not accurately represent the target population [52]. To leverage AI's profitable opportunities, we must first understand the origins of these biases. It's a common misconception that technology is neutral; in reality, ML algorithms reflect the biases of their creators and the data they are trained on [5]. AI heavily depends on human-generated data or systems created by humans, thus inheriting and amplifying human biases through complex sociotechnical systems [48]. Biases have been a part of culture and history for a long time, but the rise of digital data allows them to spread more quickly and widely [9]. There is an increasing concern that algorithms might perpetuate racial and gender disparities due to biases in their development or the data used for their training [49].

Bias in ML models is influenced by various factors, including label heterogeneity, data representation, and distribution biases [20]. Distribution bias is another critical factor, where models make misjudgments due to incorrect shortcuts derived from skewed data distributions [71]. On-device ML presents unique challenges due to hardware constraints like limited memory, computing, and energy resources, which can propagate bias through design choices [26]. Data representation bias emerges from the methods and origins of data collection. This bias can be historical, cognitive, or statistical, often arising from selection bias when sampling methods reach only a portion of the population or when the population of interest differs from the one used during model training [61]. The propagation of attribute distribution in network structures can also introduce bias; even unbiased original attributes can become biased through information propagation in a biased network structure [20].

Edge bias, unique to FL, emerges from the characteristics of each edge's dataset, leading to overfitting and incompatibility with other edge's knowledge.

This results in skewed learning in the central model and bias in the knowledge distillation process when the teacher model at the edge is overfitted or too different from the student model at the core [34]. Traditional sources of bias identified in centralized ML, such as prejudice, underestimation, and negative legacy, are also prevalent in FL. In FL, each party's local training introduces its biases into the global model through shared updates. The interactions between parties and the aggregator during training further influence the fairness of the final model [1]. Biases in AI decision-making processes are often inherited due to the closed-world assumption used in knowledge representation within datasets [32]. FL further complicates bias issues, as each party introduces its biases into the global model through shared updates. These interactions between parties and aggregators during training affect the fairness of the final model [1].

8.2.4 Bias Impact on Edge Computing Algorithms

Edge-cloud computing, FL, and other distributed computing approaches have become essential for scaling applications like object detection. These technologies help address challenges related to computation, energy consumption, privacy, and security. However, despite their benefits, these systems are not free from the influence of bias, which can significantly affect their performance. Bias in these systems can arise from several factors. One critical issue is the disparity in data classes, imbalanced labeling, and biased ground truth labels. These problems are often overlooked but can lead to inaccurate object detection, especially in environments where vehicles, edge devices, and cloud systems work together [32]. Data used in these systems can also be prone to bias because different devices—such as vehicles, sensors, and third-party applications—generate and process data differently. Environmental factors further contribute to this problem, leading to inconsistencies in data quality and biased outcomes.

Another significant issue is the bias that accumulates during the dynamic collection and scheduling of data. As data samples are collected over time, shifts in their distribution can introduce errors. This bias accumulation is particularly challenging in scenarios requiring precise perception, such as disaster response, where accuracy is crucial [71]. The homophily of sensitive attributes, which refers to the tendency of similar data points to be grouped together, can influence predictions and lead to inequalities. This is a significant issue in systems where fairness is a key consideration [63]. Moreover, FL, a common approach in edge computing, faces a global bias problem. This issue arises because wireless channels used to transmit local and global parameters can be unreliable. Poor wireless conditions may result in the loss of critical data transmissions, which biases the global model toward devices with better connectivity. As a result, the global model may drift away from the optimal solution, leading to reduced accuracy and slower learning processes [73].

8.2.5 Bias Mitigation Techniques

A variety of methods have been proposed to evaluate algorithms, including model interpretability, audits, expert analysis, and reverse engineering [65]. Algorithmic fairness also presents numerous legal challenges and complexities, with law and regulation still in the early stages in this rapidly evolving field [65]. Techniques from ML, developed to handle missing data, can help mitigate potential biases. Additionally, systems designed to identify erroneous documentation are crucial to reduce misclassification based on implicit bias [21]. It is essential to adopt procedural approaches to ensure transparency, explainability, accountability, fairness, and ethical considerations throughout the lifecycle of ML models [5]. One strategy to combat presentation bias is to incorporate elements such as diversity, novelty, and serendipity to the algorithmic process [8].

Addressing societal biases may necessitate adjusting data collection processes or manually integrating an understanding of these biases into the model-building process, and in some cases, these issues may not have technical solutions at all [42]. Numerous studies, particularly in computer science, have indicated that simply excluding race from the predictor is insufficient since protected features can be reconstructed from other variables. To address this "reconstruction problem," methods such as preprocessing data to orthogonalize explanatory variables or outcomes to race, or modifying the loss function to penalize race disparities, have been proposed [33].

8.2.5.1 Bias Mitigation in Algorithms

As concerns rise about biases in datasets and ML models, there is an essential need to implement bias mitigation algorithms. These algorithms are crafted to minimize undesirable biases, thereby ensuring that ML models are fair and trustworthy [12]. To enhance algorithmic fairness, protected attributes such as gender or ethnicity can be incorporated during the training phase to ensure that algorithmic predictions are statistically independent of these attributes. Alternatively, loss functions can be tailored for each protected group, ensuring that no single group is systematically misclassified by setting a threshold for these functions [47]. Approaches to enhancing fairness are generally grouped into three categories: preprocessing data, enforcing fairness during model training (also referred to as in-processing), and modifying outputs after the model has processed the data [53].

8.2.5.2 Preprocessing and In-processing

Preprocessing methods aim to enhance fairness in training data, thereby reducing the likelihood of producing discriminatory models. These approaches operate before the model is trained, focusing on ensuring that the data is less biased.

In the preprocessing phase, various strategies are employed to mitigate bias. Some popular ones are listed below.

- **Reweighing**: This technique adjusts the importance of training data by assigning specific weights to different combinations of groups and outcomes. The goal is to promote fairness by balancing the representation of these groups before learning.
- **Suppression**: This method involves eliminating sensitive attributes from the dataset to prevent the model from learning biases associated with these attributes.
- **Massaging**: In this approach, labels in the dataset are altered to reduce bias, ensuring that the data fed into the model is less likely to reinforce existing prejudices.
- **Multiple imputations**: This technique addresses bias introduced by missing data by replacing missing values with estimated values, thereby reducing potential skew in the model.

8.2.5.3 In-processing

In the model development stage, also known as the in-processing phase, in-processing methods are employed to address bias by integrating considerations of discriminatory behavior directly into the design and selection of the model. In the following text, we introduce several popular in-processing approaches.

- **Fairness constraints**: Incorporate fairness constraints directly into the model's objective function [48]. These constraints can ensure that the model's performance is balanced across different demographic groups.
- **Regularization techniques**: Regularization is used to penalize the model for making predictions that are biased or discriminatory. By adding a fairness regularizer to the loss function, the model is encouraged to make predictions that are both accurate and fair, reducing the risk of bias.
- **Algorithm choice**: Some algorithms are more prone to bias than others. For instance, certain decision trees might propagate biases present in the data. Explore and choose algorithms that are less likely to amplify biases.
- **Adversarial debiasing**: Implement adversarial techniques where a secondary model (the adversary) is trained to predict the sensitive attributes from the main model's predictions. The main model is then adjusted to minimize the adversary's accuracy, thereby reducing bias.
- **Fair representation learning**: Fair representation learning is a technique that develops a model to make accurate predictions while eliminating biases associated with sensitive information [50]. In this approach, the model is designed to learn representations of the data that are invariant to sensitive attributes, such as race or gender.

These in-processing approaches are crucial for mitigating bias during model development, ensuring that the resulting model is not only effective but also equitable in its decision-making process. These approaches can be combined to build more robust and equitable models. For instance, the integration of fair representation learning with adversarial learning strategies may create models where the distribution of data representations, conditioned on each sensitive attribute, remains uniform across different groups. This helps to address both direct and indirect biases, leading to more fair and balanced outcomes in various applications [66].

8.2.5.4 Postprocessing

Postprocessing techniques focus on adjusting model outputs to achieve fairness [48]. In the postprocessing stage of an algorithm, the aim is to modify the results to ensure fairness. The ideal postprocessing debiasing approach considers both group and individual fairness. However, addressing bias and maintaining accuracy can be challenging, as they often conflict. Specifically, focusing on individual bias reduction in postprocessing might lead to a decrease in accuracy [36]. Various postprocessing bias mitigation techniques and performance metrics have been proposed to adjust and reduce biases within the prediction models [45].

8.2.6 Open Research Problems

Research in bias mitigation is ongoing, with significant advancements in techniques aimed at enhancing fairness and minimizing indirect prejudices stemming from algorithmic predictions [21]. Policymakers are confronted with difficult and impactful decisions, often facing uncertainty about the most appropriate course of action in various contexts [18]. As methods to debias ML algorithms continue to develop, there are also improvements in techniques to enhance fairness and reduce indirect prejudices that result from algorithm predictions [21]. ML techniques designed to handle missing data can help control for potential biases.

It is imperative to consider, adopt and institute procedural approaches to ensure transparency, explainability, accountability, fairness and ethical considerations of the underlying models used in lifecycles of ML applications [5]. Despite these advances, numerous research problems remain unexplored, including bias mitigation in multimodal data, the development of novel evaluation metrics for bias, addressing bias in generative models, improving interpretability about bias in model, and achieving fairness in dynamic environments.

8.3 Summary and Practice

8.3.1 Summary

This chapter focused on two important ethical and social concerns: privacy and bias.

In the section on privacy, we began by discussing the importance of privacy, defining various forms of privacy, and outlining key privacy concerns. We then introduced a suite of major privacy-preserving techniques, including traditional anonymization approaches, k-anonymity, differential privacy, homomorphic encryption, and FL. For each of these techniques, we provided examples of their application and implementation within edge computing environments.

Regarding bias, we analyzed different types of biases and their underlying causes. Additionally, we explored how these biases impact the performance of edge computing algorithms, especially those machine learning and deep learning algorithms. Moreover, we presented a series of biases mitigation strategies designed to address these challenges.

Despite the extensive efforts by researchers to tackle privacy and bias issues, significant open research challenges remain, which we have summarized at the conclusion of each topic.

8.3.2 Practice Questions

1) Identify the key factors that contribute to biases in AI models deployed on edge devices and discuss how these biases might manifest in real-world applications.
2) Discuss the potential privacy concerns associated with collecting and processing data at the edge.
3) Analyze how biases in edge computing algorithms can impact different user groups, and outline steps to mitigate these biases.
4) Examine the legal measures necessary to ensure privacy protection in edge computing.
5) Discuss the technical challenges of implementing privacy-preserving techniques such as differential privacy and federated learning on edge devices, and how do these challenges affect privacy and bias.

8.3.3 Course Projects

1) Evaluate the privacy concerns of a basic edge computing application.
2) Implement and evaluate a privacy-preserving algorithm.
3) Assess bias in a given dataset.
4) Implement and evaluate a bias mitigation algorithm.
5) Propose new strategies to minimize biases in a specific edge computing algorithm.
6) Design protocol for secure data aggregation in edge networks.
7) Create a framework for auditing decisions made by AI models on edge devices, ensuring transparency and accountability in AI-driven processes.
8) Implement edge computing for privacy-preserving applications in important domains such as healthcare, connected autonomous vehicles, and others.

9) Evaluate the potential of blockchain technology in enhancing privacy and reducing bias in edge computing applications.
10) Design a system for real-time privacy monitoring in edge IoT applications.

Chapter 8 Suggested Papers

1 Nicol Turner Lee, Paul Resnick, and Genie Barton. "Algorithmic bias detection and mitigation: Best practices and policies to reduce consumer harms". In: *Brookings Institute*: Washington, DC, USA 2 (2019).

2 Pasika Ranaweera, Anca Delia Jurcut, and Madhusanka Liyanage. "Survey on multi-access edge computing security and privacy". In: *IEEE Communications Surveys & Tutorials* 23. 2 (2021), pp. 1078–1124.

3 Mark Ryan. "The future of transportation: Ethical, legal, social and economic impacts of self-driving vehicles in the year 2025". In: *Science and Engineering Ethics* 26. 3 (2020), pp. 1185–1208.

4 Kewei Sha et al. "On security challenges and open issues in Internet of Things". In: *Future Generation Computer Systems* 83 (2018), pp. 326–337.

5 Kewei Sha et al. "A survey of edge computing-based designs for IoT security". In: *Digital Communications and Networks* 6. 2 (2020), pp. 195–202.

References

1 Annie Abay et al. "Mitigating bias in federated learning". In: *arXiv preprint arXiv:2012.02447* (2020).

2 Haftay Gebreslasie Abreha, Mohammad Hayajneh, and Mohamed Adel Serhani. "Federated learning in edge computing: A systematic survey". In: *Sensors* 22. 2 (2022), p. 450.

3 Ibrahim Bio Abubakar, Tarjana Yagnik, and Kabiru Mohammed. "Robustness of k-anonymization model in compliance with general data protection regulation". In: *2022 5th International Conference on Computing and Big Data (ICCBD)*. IEEE. 2022, pp. 67–72.

4 Abbas Acar et al. "A survey on homomorphic encryption schemes: Theory and implementation". In: *ACM Computing Surveys (CSUR)* 51. 4 (2018), pp. 1–35.

5 Shahriar Akter et al. Algorithmic bias in data-driven innovation in the age of AI. *International Journal of Information Management* 60 (2021), p. 102387.

6 Shahriar Akter et al. "Algorithmic bias in machine learning-based marketing models". In: *Journal of Business Research* 144 (2022), pp. 201–216.

7 Abdulmalik Alwarafy et al. "A survey on security and privacy issues in edge-computing-assisted Internet of Things". In: *IEEE Internet of Things Journal* 8. 6 (2020), pp. 4004–4022.

8 Ricardo Baeza-Yates. "Data and algorithmic bias in the web". In: *Proceedings of the 8th ACM Conference on Web Science*. 2016, pp. 1–1.

9 Ricardo Baeza-Yates. "Bias on the web". In: *Communications of the ACM* 61. 6 (2018), pp. 54–61.

10 Tianyu Bai, Qing Yang, and Song Fu. "User-defined privacy preserving data sharing for connected autonomous vehicles utilizing edge computing". In: *2023 IEEE/ACM Symposium on Edge Computing (SEC)*. IEEE. 2023, pp. 145–157.

11 Roberto J Bayardo and Rakesh Agrawal. "Data privacy through optimal k-anonymization". In: *21st International Conference on Data Engineering (ICDE'05)*. IEEE. 2005, pp. 217–228.

12 Karan Bhanot et al. "Stress-testing bias mitigation algorithms to understand fairness vulnerabilities". In: *Proceedings of the 2023 AAAI/ACM Conference on AI, Ethics, and Society*. 2023, pp. 764–774.

13 Mengnan Bi et al. "A privacy-preserving mechanism based on local differential privacy in edge computing". In: *China Communications* 17. 9 (2020), pp. 50–65.

14 Su Lin Blodgett et al. "Language (technology) is power: A critical survey of" bias" in NLP". In: *arXiv preprint arXiv:2005.14050* (2020).

15 Engin Bozdag. "Bias in algorithmic filtering and personalization". In: *Ethics and Information Technology* 15 (2013), pp. 209–227.

16 Alexander Brecko et al. "Federated learning for edge computing: A survey". In: *Applied Sciences* 12. 18 (2022), p. 9124.

17 Wei Chang and Jie Wu. *Fog/Edge Computing for Security, Privacy, and Applications*. Springer, 2021.

18 Sam Corbett-Davies et al. "Algorithmic decision making and the cost of fairness". In: *Proceedings of the 23rd ACM SIGKDD International Conference on Knowledge Discovery and Data Mining*. 2017, pp. 797–806.

19 David Danks and Alex John London. "Algorithmic bias in autonomous systems". In: *Ijcai* 17. 2017 (2017), pp. 4691–4697.

20 Yushun Dong et al. "EDITS: Modeling and mitigating data bias for graph neural networks". In: *Proceedings of the ACM Web Conference 2022*. 2022, pp. 1259–1269.

21 Milena A Gianfrancesco et al. "Potential biases in machine learning algorithms using electronic health record data". In: *JAMA Internal Medicine* 178. 11 (2018), pp. 1544–1547.

22 Sara Hajian, Francesco Bonchi, and Carlos Castillo. "Algorithmic bias: From discrimination discovery to fairness-aware data mining". In: *Proceedings of the 22nd ACM SIGKDD International Conference on Knowledge Discovery and Data Mining*. 2016, pp. 2125–2126.

23 Muneeb Ul Hassan, Mubashir Husain Rehmani, and Jinjun Chen. "Differential privacy techniques for cyber physical systems: A survey". In: *IEEE Communications Surveys & Tutorials* 22. 1 (2019), pp. 746–789.

24 Anna Lauren Hoffmann. "Where fairness fails: Data, algorithms, and the limits of antidiscrimination discourse". In: *Information, Communication & Society* 22. 7 (2019), pp. 900–915.

25 Chris Jay Hoofnagle, Bart Van Der Sloot, and Frederik Zuiderveen Borgesius. "The European Union general data protection regulation: What it is and what it means". In: *Information & Communications Technology Law* 28. 1 (2019), pp. 65–98.

26 Wiebke Hutiri et al. "Tiny, always-on, and fragile: Bias propagation through design choices in on-device machine learning workflows". In: *ACM Transactions on Software Engineering and Methodology* 32. 6 (2023), pp. 1–37.

27 Priyank Jain, Manasi Gyanchandani, and Nilay Khare. "Big data privacy: A technological perspective and review". In: *Journal of Big Data* 3 (2016), pp. 1–25.

28 Zhanglong Ji, Zachary C Lipton, and Charles Elkan. "Differential privacy and machine learning: A survey and review". In: *arXiv preprint arXiv:1412.7584* (2014).

29 Tao Jiang and Yubo Song. "Privacy protection offloading algorithm based on K-anonymity in mobile edge computing networks". In: *2023 IEEE 11th International Conference on Information, Communication and Networks (ICICN)*. IEEE. 2023, pp. 248–254.

30 Pengfei Jiang and Lei Ying. "An optimal stopping approach for iterative training in federated learning". In: *2020 54th Annual Conference on Information Sciences and Systems (CISS)*. IEEE. 2020, pp. 1–6.

31 Bin Jiang et al. "Privacy-preserving federated learning for industrial edge computing via hybrid differential privacy and adaptive compression". In: *IEEE Transactions on Industrial Informatics* 19. 2 (2021), pp. 1136–1144.

32 Dewant Katare et al. "Bias detection and generalization in AI algorithms on edge for autonomous driving". In: *2022 IEEE/ACM 7th Symposium on Edge Computing (SEC)*. IEEE. 2022, pp. 342–348.

33 Jon Kleinberg et al. "Algorithmic fairness". In: *AEA Papers and Proceedings*. Vol. 108. American Economic Association 2014 Broadway, Suite 305, Nashville, TN 37203. 2018, pp. 22–27.

34 Sangho Lee, Kiyoon Yoo, and Nojun Kwak. "Edge bias in federated learning and its solution by buffered knowledge distillation". In: *arXiv preprint arXiv:2010.10338* (2020).

35 Kristen LeFevre, David J DeWitt, and Raghu Ramakrishnan. "Incognito: Efficient full-domain k-anonymity". In: *Proceedings of the 2005 ACM SIGMOD International Conference on Management of Data*. 2005, pp. 49–60.

36 Pranay Lohia. "Priority-based post-processing bias mitigation for individual and group fairness". In: *arXiv preprint arXiv:2102.00417* (2021).

37 Waranya Mahanan, W Art Chaovalitwongse, and Juggapong Natwichai. “Data anonymization: A novel optimal k-anonymity algorithm for identical generalization hierarchy data in IoT”. In: *Service Oriented Computing and Applications* 14 (2020), pp. 89–100.

38 Abdul Majeed and Sungchang Lee. “Anonymization techniques for privacy preserving data publishing: A comprehensive survey”. In: *IEEE Access* 9 (2020), pp. 8512–8545.

39 Chiara Marcolla et al. “Survey on fully homomorphic encryption, theory, and applications”. In: *Proceedings of the IEEE* 110. 10 (2022), pp. 1572–1609.

40 Joana Ferreira Marques and Jorge Bernardino. “Analysis of data anonymization techniques”. In: *KEOD*. 2020, pp. 235–241.

41 Ninareh Mehrabi et al. “A survey on bias and fairness in machine learning”. In: *ACM Computing Surveys (CSUR)* 54. 6 (2021), pp. 1–35.

42 Shira Mitchell et al. “Algorithmic fairness: Choices, assumptions, and definitions”. In: *Annual Review of Statistics and its Application* 8. 1 (2021), pp. 141–163.

43 Kundan Munjal and Rekha Bhatia. “A systematic review of homomorphic encryption and its contributions in healthcare industry”. In: *Complex & Intelligent Systems* 9. 4 (2023), pp. 3759–3786.

44 Suntherasvaran Murthy et al. “A comparative study of data anonymization techniques”. In: *2019 IEEE 5th International Conference on Big Data Security on Cloud (BigDataSecurity), IEEE Intl Conference on High Performance and Smart Computing, (HPSC) and IEEE Intl Conference on Intelligent Data and Security (IDS)*. IEEE. 2019, pp. 306–309.

45 Lama H Nazer et al. “Bias in artificial intelligence algorithms and recommendations for mitigation”. In: *PLOS Digital Health* 2. 6 (2023), e0000278.

46 Dinh C Nguyen et al. “Federated learning meets blockchain in edge computing: Opportunities and challenges”. In: *IEEE Internet of Things Journal* 8. 16 (2021), pp. 12806–12825.

47 Natalia Norori et al. “Addressing bias in big data and AI for health care: A call for open science”. In: *Patterns* 2. 10 (2021), 100347.

48 Eirini Ntoutsi et al. “Bias in data-driven artificial intelligence systems—An introductory survey”. In: *Wiley Interdisciplinary Reviews: Data Mining and Knowledge Discovery* 10. 3 (2020), e1356.

49 Ziad Obermeyer et al. “Dissecting racial bias in an algorithm used to manage the health of populations”. In: *Science* 366. 6464 (2019), pp. 447–453.

50 Changdae Oh et al. “Learning fair representation via distributional contrastive disentanglement”. In: *Proceedings of the 28th ACM SIGKDD Conference on Knowledge Discovery and Data Mining*. 2022, pp. 1295–1305.

51 Ravi B Parikh, Stephanie Teeple, and Amol S Navathe. “Addressing bias in artificial intelligence in health care”. In: *Jama* 322. 24 (2019), pp. 2377–2378.

52 Dana Pessach and Erez Shmueli. "Algorithmic fairness". In: *Machine Learning for Data Science Handbook: Data Mining and Knowledge Discovery Handbook*. Ed. by Lior Rokach, Oded Maimon, and Erez Shmueli. Springer, 2023, pp. 867–886.

53 Felix Petersen et al. "Post-processing for individual fairness". In: *Advances in Neural Information Processing Systems* 34 (2021), pp. 25944–25955.

54 Pasika Ranaweera, Anca Delia Jurcut, and Madhusanka Liyanage. "Survey on multi-access edge computing security and privacy". In: *IEEE Communications Surveys & Tutorials* 23. 2 (2021), pp. 1078–1124.

55 Wang Ren et al. "Privacy enhancing techniques in the Internet of Things using data anonymisation". In: *Information Systems Frontiers* (2021), pp. 1–12.

56 Wilson Santos et al. "Data anonymization: K-anonymity sensitivity analysis". In: *2020 15th Iberian Conference on Information Systems and Technologies (CISTI)*. IEEE. 2020, pp. 1–6.

57 Rathindra Sarathy and Krishnamurty Muralidhar. "Evaluating Laplace noise addition to satisfy differential privacy for numeric data". In: *Transactions on Data Privacy* 4. 1 (2011), pp. 1–17.

58 Khotso Selialia, Yasra Chandio, and Fatima M Anwar. "Mitigating group bias in federated learning for heterogeneous devices". In: *The 2024 ACM Conference on Fairness, Accountability, and Transparency*. 2024, pp. 1043–1054.

59 Kewei Sha et al. "On security challenges and open issues in Internet of Things". In: *Future Generation Computer Systems* 83 (2018), pp. 326–337.

60 Kewei Sha et al. "A survey of edge computing-based designs for IoT security". In: *Digital Communications and Networks* 6. 2 (2020), pp. 195–202.

61 Nima Shahbazi et al. "Representation bias in data: A survey on identification and resolution techniques". In: *ACM Computing Surveys* 55. 13s (2023), pp. 1–39.

62 Swapnil Sadashiv Shinde et al. "On the design of federated learning in latency and energy constrained computation offloading operations in vehicular edge computing systems". In: *IEEE Transactions on Vehicular Technology* 71. 2 (2021), pp. 2041–2057.

63 Indro Spinelli et al. "FairDrop: Biased edge dropout for enhancing fairness in graph representation learning". In: *IEEE Transactions on Artificial Intelligence* 3. 3 (2021), pp. 344–354.

64 Frank Stinar, Zihan Xiong, and Nigel Bosch. "An approach to improve k-anonymization practices in educational data mining". In: *Journal of Educational Data Mining* 16. 1 (2024), pp. 61–83.

65 Allison Woodruff et al. "A qualitative exploration of perceptions of algorithmic fairness". In: *Proceedings of the 2018 Chi Conference on Human Factors in Computing Systems*. 2018, pp. 1–14.

66 Le Wu et al. "Learning fair representations for recommendation: A graph-based perspective". In: *Proceedings of the Web Conference 2021*. 2021, pp. 2198–2208.

67 Qi Xia et al. "A survey of federated learning for edge computing: Research problems and solutions". In: *High-Confidence Computing* 1. 1 (2021), p. 100008.

68 Xiaoyan Yan, Qilin Wu, and Youming Sun. "A homomorphic encryption and privacy protection method based on blockchain and edge computing". In: *Wireless Communications and Mobile Computing* 2020. 1 (2020), p. 8832341.

69 Aiting Yao et al. "Differential privacy in edge computing-based smart city Applications: Security issues, solutions and future directions". In: *Array* 19 (2023), p. 100293.

70 Yunfan Ye et al. "EdgeFed: Optimized federated learning based on edge computing". In: *IEEE Access* 8 (2020), pp. 209191–209198.

71 Tiankuo Yu et al. "Bias-compensation augmentation learning for semantic segmentation in UAV networks". In: *IEEE Internet of Things Journal* 11. 12 (2024), pp. 21261–21273.

72 Muhammad Bilal Zafar et al. "Fairness constraints: Mechanisms for fair classification". In: *Artificial Intelligence and Statistics*. PMLR. 2017, pp. 962–970.

73 Ruslan Zhagypar et al. "Characterization of the global bias problem in aerial federated learning". In: *IEEE Wireless Communications Letters* 12. 8 (2023), pp. 1339–1343.

74 Xuejun Zhang et al. "Differential privacy-based indoor localization privacy protection in edge computing". In: *2019 IEEE SmartWorld, Ubiquitous Intelligence & Computing, Advanced & Trusted Computing, Scalable Computing & Communications, Cloud & Big Data Computing, Internet of People and Smart City Innovation (SmartWorld/SCALCOM/UIC/ATC/CBDCom/IOP/SCI)*. IEEE. 2019, pp. 491–496.

75 Lejun Zhang et al. "Secure and efficient data storage and sharing scheme for blockchain-based mobile-edge computing". In: *Transactions on Emerging Telecommunications Technologies* 32. 10 (2021), e4315.

76 Shiwen Zhang et al. "A caching-based dual k-anonymous location privacy-preserving scheme for edge computing". In: *IEEE Internet of Things Journal* 10. 11 (2023), pp. 9768–9781.

9

Conclusion and Future Directions

9.1 Key Insights and Conclusions

The rapid growth of applications and services in the Internet of Everything (IoE) has given rise to edge computing, establishing it as a critical hardware and software platform in the era of edge-based big data processing. In the IoE era, the enormous volumes of data generated by edge devices, coupled with the demand for real-time processing, have introduced new challenges related to transmission bandwidth, computational load, and data privacy under the traditional cloud computing model. Edge computing addresses these challenges by introducing a decentralized data-processing model, which not only reduces response times and improves reliability but also enables more applications to migrate from cloud computing centers to network edge devices through context-aware technologies. Processing data at the edge, without uploading it to the cloud, significantly reduces transmission bandwidth requirements, lowers energy consumption for data transmission, alleviates the load on cloud centers, and enhances data privacy.

This book has examined edge computing from multiple perspectives, including its necessity, fundamental principles, architecture and components, edge intelligence, challenges and opportunities, future trends, practical applications, and privacy and bias concerns. A key takeaway is that edge computing and cloud computing are not mutually exclusive but rather complementary. The edge computing model still supports traditional cloud computing while also enabling remote access to computing resources, facilitating data sharing and collaboration. By decentralizing computing power from centralized cloud infrastructures to edge devices, edge computing provides significant benefits, such as reduced latency and enhanced privacy.

As artificial intelligence (AI) rapidly evolves, the integration of edge computing with AI has given rise to edge intelligence, laying the foundation for efficient machine learning and the broader adoption of AI technologies. This book

Edge Computing: Systems and Applications, First Edition. Lanyu Xu and Weisong Shi.

Companion Website: www.wiley.com/go/xu/computing

has highlighted the importance of edge intelligence, focusing on the available hardware and software support, the enabling technologies, and the strategies for designing these systems.

To accelerate the deployment and adoption of edge computing, this book identifies several key issues that are urgently needed to be addressed in current research. These issues are multifaceted, including programmability, data management, resource allocation, and security. These challenges are critical barriers to the seamless integration and widespread adoption of edge computing technologies. It also discussed the importance of developing robust deployment strategies and business models to sustain the growth of edge computing.

With a forward-looking perspective, we emphasized the importance of staying ahead of technological advancements and exploring new paradigms like sky computing and meta computing. We also identified emerging trends, such as the integration of edge computing with AI, the potential impact of 6G, and the exciting possibilities of edge computing in space exploration.

Through various studies, we demonstrated the transformative impact of edge computing across industries like manufacturing, telecommunications, healthcare, smart cities, the Internet of Things, retail, and autonomous vehicles. These case studies provide concrete evidence of how edge computing is being applied to solve real-world problems while also highlighting the challenges and lessons learned from these implementations, offering system prototypes that can serve as valuable references for further research.

As with any new information technology, its acceptance by the industry hinges on ensuring safety and reliability. Similarly, the security of edge computing is a major concern for both academia and industry. Before edge computing can be effectively applied in smart cities, intelligent transportation, smart homes, smart manufacturing, or collaborative platforms, security and privacy protection issues must be addressed. This book emphasized the need for robust privacy protection mechanisms and highlighted the risks associated with the digital divide, with the hope of stimulating further research and development in the field of edge computing security and privacy.

9.2 So, What Is Next?

Building on the key insights gathered earlier, we conclude this book with recommendations for future research to address the challenges and opportunities within edge computing.

In Chapter 6, we explored the future trends of edge computing, highlighting the promising directions this field is heading toward. However, as our case studies in Chapter 7 illustrate, the integration of edge computing across various societal domains, such as healthcare, smart cities, and urban planning, remains a complex endeavor. Significant research and development efforts are still needed to fully leverage the potential of edge computing. The challenges identified in Chapter 5 underscore the need for innovative solutions to seamlessly integrate edge computing into various industries.

Despite nearly a decade of development, edge computing is at a critical juncture where its role is becoming increasingly vital, especially as AI models grow in complexity and size. The dominance of large AI models by a few leading enterprises has created a significant barrier to their deployment on resource-constrained edge devices, limiting their widespread adoption.

From a model perspective, enabling powerful AI models on edge devices with limited resources is, and will continue to be, a major research focus. In Chapter 4, we thoroughly reviewed current model compression techniques, such as quantization, pruning, and knowledge distillation. These methods are essential for making advanced AI models feasible on edge devices. However, as AI models become more sophisticated and complicated, there is an ongoing need to develop even more effective compression techniques that maintain high performance while minimizing computational demands.

From a system perspective, maximizing the potential of edge computing remains an urgent priority. In Chapter 3, we discussed several open problems that future researchers should consider, not only when exploring existing edge-cloud collaborations but also when investigating emerging technologies.

We are open to embracing emerging technologies and would eager to see how new technologies integrate with edge computing to guide the next phase of innovation in this field. Although quantum computing is still in its early stages and is not within the scope of this book, it holds the potential to revolutionize edge computing by providing unparalleled computational power, enabling edge devices to perform tasks that are currently impossible due to resource constraints.

As a computing paradigm intended for everyday use, privacy and security will always be the top concerns in edge computing. Future research should prioritize the development of advanced security frameworks, such as lightweight encryption techniques, secure data-sharing protocols, and optimized privacy-preserving AI models. Although not covered in this book, blockchain technology can be explored as a potential solution for enhancing security and trustworthiness in decentralized edge networks. We must ensure that edge computing is developed in a way that is safe and secure, ethical, and inclusive.

These recommendations are intended to guide the next wave of research and innovation in edge computing, addressing both the challenges and the opportunities that lie ahead. By advancing in these areas, the edge computing community can continue to push the boundaries of what is possible, leading to more powerful, efficient, and equitable systems that will play a pivotal role in the future of technology.

We hope this book will inspire further development in edge computing across both industry and academia, attracting a growing number of researchers and practitioners from fields such as computer science and engineering to delve deeper into this area. By leveraging the wealth of knowledge accumulated in current information technology, we aim to identify and address the remaining scientific and engineering challenges in edge computing research.

Collaboration beyond traditional technical fields—such as computer systems, communications, networks, and applications—is essential to further promote the realization and adoption of edge computing frameworks and models. It is equally important to involve professionals and research institutions from other related fields, such as environmental science, public health, law enforcement, firefighting, and public services, to foster interdisciplinary collaboration between computer scientists and experts from various industries. This collaboration will be crucial in solving real-world problems and exploring new frontiers, ensuring that the benefits of edge computing extend across all aspects of society.

Edge computing represents a collaborative computing model that integrates diverse resources, offering an unprecedented opportunity in the history of computer science. Over the past decade, edge computing has made significant strides, and we are confident that the next decade will see even remarkable developments in edge computing and edge intelligence. As we enter the IoE era, the synergy between AI and edge computing will drive a new revolution in information technology, propelling humanity into a new era of technological advancement.

Index

a

AI models
  federated learning 63
  inference on edge 107
  training on edge 104–105
algorithms
  load balancing 133
  optimization algorithms 176–177
applications
  autonomous vehicles 194, 213–217
  healthcare 200–203
  industrial IoT 210–211
  smart cities 203–209
architecture
  edge computing architecture 47–49, 53–54, 71, 146, 197, 199, 210
  network architecture 10–11, 30, 61, 68–69, 79, 84, 175, 177–178, 186
  system architecture 12, 139, 183
artificial intelligence (AI)
  deep learning 40, 60, 79, 83, 85–88, 91, 94, 96, 101–104, 108, 111, 144, 147, 198, 211, 223, 246
  edge AI 17, 63, 70, 77–80, 91–111, 144
  machine learning 14, 17, 26, 51, 59, 63, 70, 77–78, 84–88, 90, 104, 108, 142, 144, 198–199, 202–203, 214, 223, 246, 253

b

background of edge computing 15
bandwidth
  latency and bandwidth 10, 37, 57, 63, 142–143, 171, 200, 202–203, 205–207, 210–211
  network bandwidth 3, 5, 27, 36, 48, 77, 102, 142
bias 223–246, 253
big data
  data analytics 61
  data processing 3, 4, 8, 18, 23, 26, 35–36, 39–40, 42, 253
  data storage 8, 31, 41, 51, 59, 68, 129, 138–139, 143, 145, 159, 193, 203, 225, 233
blockchain 138–140, 151, 202, 210, 255

c

case studies
  autonomous vehicles 213–217
  healthcare applications 200–203
  smart cities 203–209
challenges
  in edge computing 123–151, 256
  latency challenges 214
  resource management 185–186
  security challenges 137

Edge Computing: Systems and Applications, First Edition. Lanyu Xu and Weisong Shi.

Companion Website: www.wiley.com/go/xu/computing

cloud computing
cloud data centers 16, 39, 53, 65, 143, 182
cloud integration 70, 158
edge *vs.* cloud 33–41
collaborative models 60
communication
5G networks 30, 57, 69, 174–175, 194
low-latency communication 182
compression
model compression 94, 100, 108, 111–112, 172, 255
computational power
energy efficiency 6, 69, 85, 104, 107, 148, 194, 203–204, 209
resource allocation 89, 130–136, 214, 216, 254

d

data
data privacy 6, 8, 36, 50, 61, 127–128, 172, 202–203, 205–207, 213, 215, 225–226, 228–230, 233–235, 253
data security 6, 28, 36, 41, 49, 62, 136, 138–139, 194, 200, 202–203, 206, 224
real-time data 1, 3–4, 31, 40, 48–51, 58, 61–63, 86, 88, 90, 130, 138, 171, 178, 193–198, 200, 202–204, 206–207, 210–211, 214, 216–217
deep learning
model training 88
neural networks 82, 92, 94, 107, 144, 165, 234
devices
edge devices 3–4, 8, 9–11, 14, 17, 23, 27–32, 37, 39–40, 47–48, 54–55, 63–65, 70, 78, 81–83, 85, 87–90, 97, 100–102, 104–108, 111–112, 124–129, 131–133, 135–138, 143–145, 148, 151, 166–173, 200, 202–203, 206, 210, 214, 225, 234, 236, 242, 253, 255
IoT devices 4, 15, 47, 82, 87, 128, 140, 146, 166, 194, 200, 204–208, 210–211
mobile devices 5, 10–11, 15, 35, 55, 82, 84–85, 103
digital divide 236–245, 254
distributed computing
distributed network 5, 47, 68–69
distributed storage 9, 139, 148
distributed systems 9, 25–26, 66, 130, 140, 147–148
federated learning 63, 105, 168, 213, 234–236

e

edge computing
applications 193–217
architecture 47–49, 53–54, 71, 146, 197, 199, 210
challenges 39–41, 123–151, 256
environments 2, 6, 8, 10–11, 24–25, 31–33, 52–53, 55, 57, 61–63, 65–67, 69–71, 79, 83, 87–88, 90–91, 102, 104, 107, 111, 125–127, 130–131, 133, 136, 138–141, 143–146, 149, 159, 166, 168, 171, 178, 183, 186, 195–197, 203–208, 210–212, 216, 224–225, 227, 236, 242, 245–246
future directions 206, 253–256
in healthcare 201
integration 1–2, 12, 14, 30–31, 35, 40, 42, 53, 57, 59, 65, 70–71, 125, 138–139, 142–147, 151, 158–159, 161, 164–174, 179–180, 200–212, 244, 253–255
in manufacturing 196

models 7–10, 55–65, 147–149
privacy 6, 8, 36, 50, 61, 127–128, 136, 172, 202–203, 205–207, 213, 215, 225–226, 229–230, 233–235, 253
security 28, 136
in smart cities 204
trends 12, 17, 70, 157–186, 194, 240, 253–254
edge intelligence
knowledge distillation 14, 91, 98–100, 108, 111, 168–169, 242, 255
low-rank approximation 97–98
pruning 85, 91–94, 100, 111, 144, 169, 255
quantization 87, 94–97, 100, 106, 108, 144, 169–170, 172, 255
emerging paradigms 157–186
energy efficiency
power consumption 81

f

fog computing 7, 11–12, 203–205, 209
fundamental principles 23–42, 253
future directions 206, 253–256

h

hardware
edge devices 81–86
hardware-software codesign 101–102
healthcare
predictive maintenance 200, 203
remote patient monitoring 194, 200, 203

i

infrastructure 3, 6, 11, 13, 34, 41, 47–55, 65–69, 123, 136, 142–143, 146, 149, 151, 158, 175, 183, 198–200, 203, 206–208, 215, 217
integration 1–2, 12, 14, 30–31, 35, 40, 42, 53, 57, 59, 65, 70–71, 125, 138–139, 142–147, 151, 158, 159, 161, 164–174, 179–180, 200–212, 244, 253–255
internet of things (IoT)
IoT applications 5–6, 50, 84, 210–212
IoT devices 4, 15, 47, 82, 87, 128, 140, 146, 166, 194, 200, 204–207, 210–211

l

latency
low latency 6, 14, 23, 31, 36, 47, 52, 55, 60–61, 65, 69, 103, 111, 130, 143, 146, 171–172, 182, 184–185, 194, 197–199, 201, 204–205, 207, 209–210, 213–215
load balancing
resource allocation 133
task scheduling 131

m

machine learning
edge AI 17, 63, 70, 86, 144
model compression 94, 100, 108, 111–112, 172
model inference 107
model training 104, 215, 235, 241, 243
mobile edge computing (MEC) 10–11, 55, 176
applications 176

n

networking
edge networking 65–66
5G networks 30, 57, 69, 174–175, 194
nodes
cloud nodes 37, 38

nodes (*contd.*)
edge nodes 10–11, 33, 37–38, 41, 48, 52–53, 57–58, 60–65, 68–71, 104, 124–126, 128–129, 131, 133, 142–143, 148–149, 151, 167–168, 183, 198–199, 225, 232, 236–237

o

optimization
performance optimization 90
resource optimization 130–136, 216

p

privacy
data privacy 6, 8, 36, 50, 61, 127–128, 136, 172, 202–203, 205–207, 213, 215, 225–226, 229–230, 233–235, 253
privacy concerns 139, 167, 170, 207, 223–225, 227, 245
privacy-preserving 223–224, 227–235, 246, 255
programmability 10, 54–55, 123–130, 254

r

real-time processing
real-time analytics 14, 26, 51, 200
real-world case studies 193
resource management
resource allocation 58, 60, 70, 89, 130–136, 202, 214
resource optimization 130–136, 216

s

security 136–142
in edge computing 28, 136
sky computing 53, 158–161, 163, 186, 254
smart cities
urban 195, 200, 204–209
storage
cache 52
data storage 8, 28, 31, 41, 51, 59, 129, 138–139, 143, 159, 193, 203, 225, 233

t

technologies
blockchain 138–140, 151, 202, 210, 255
emerging technologies 157–186
threat 82, 137–138, 216
trends 12, 17, 70, 157–186, 194, 240, 253–254

v

vehicles
autonomous vehicles 14, 86, 89, 137, 139, 145, 171, 194, 212–217, 224, 254
connected vehicles 49, 58, 78, 139, 214

w

wireless communication
5G networks 30, 57, 69, 174–175, 194
low-latency communication 65, 182, 185, 198, 213, 215

Printed and bound by CPI Group (UK) Ltd, Croydon, CR0 4YY

17/09/2026

14964333-0002